XXXIX Symposium on Nuclear Physics (Cocoyoc 2016)

Journal of Physics: Conference Series
Volume 730

Cocoyoc, Mexico
5 – 8 January 2016

Editors:

Luis Acosta
Roelof Bijker

ISBN: 978-1-5108-2728-8
ISSN: 1742-6588

Printed from e-media with permission by:

Curran Associates, Inc.
57 Morehouse Lane
Red Hook, NY 12571

Some format issues inherent in the e-media version may also appear in this print version.

Printed by Curran Associates, Inc. (2016)

For permission requests, please contact the Institute of Physics at the address below.

Institute of Physics
Dirac House, Temple Back
Bristol BS1 6BE UK

Phone: 44 1 17 929 7481
Fax: 44 1 17 920 0979

techtracking@iop.org

Additional copies of this publication are available from:

Curran Associates, Inc.
57 Morehouse Lane
Red Hook, NY 12571 USA
Phone: 845-758-0400
Fax: 845-758-2633
Email: curran@proceedings.com
Web: www.proceedings.com

Table of contents

Volume 730

XXXIX Symposium on Nuclear Physics 2016 (Cocoyoc2016)
5–8 January 2016, Cocoyoc, Morelos, Mexico

Accepted papers received: 8 July 2016
Published online: 1 August 2016

Preface

Papers

Journal of Physics: Conference Series **730** (2016) 011001 doi:10.1088/1742-6596/730/1/011001

39th Symposium on Nuclear Physics

Cocoyoc, Mexico 2016

XXXIX Symposium on Nuclear Physics

Hacienda Cocoyoc, Morelos, Mexico
January 5-8, 2016

EDITORS

LUIS ACOSTA
Instituto de Física
Universidad Nacional Autónoma de México
ROELOF BIJKER
Instituto de Ciencias Nucleares
Universidad Nacional Autónoma de México

International Advisory Committee

Ani Aprahamian	Notre Dame, USA
Daniel Bardayan	Notre Dame, USA
Francesco Cappuzzello	INFN-LNS, Italy
Osvaldo Civitarese	La Plata, Argentina
Jerry P. Draayer	LSU, USA
Alfredo Galindo-Uribarri	ORNL & UT, USA
María José García Borge	IEM-CSIC, Spain
Paulo Gomes	Fluminense, Brazil
Reiner Kruecken	TRIUMF, Canada
Jorge López	UTEP, USA
Takashi Nakamura	Tokyo Institute of Technology, Japan
Thomas Papenbrock	ORNL, USA
Jorge Piekarewicz	FSU, USA
Stuart Pittel	Delaware, USA
Elena Santopinto	Genova, Italy
Mike Snow	Indiana, USA
Adam Szczepaniak	Indiana, USA
Piet Van Isacker	GANIL, France
Michael Wiescher	Notre Dame, USA

Organizing Commmittee

Luis Acosta	IF-UNAM
Roelof Bijker	ICN-UNAM

Sponsors

Instituto de Ciencias Nucleares, UNAM
Instituto de Física, UNAM
Instituto Nacional de Investigaciones Nucleares
División de Física Nuclear de la SMF

Journal of Physics: Conference Series **730** (2016) 011001 doi:10.1088/1742-6596/730/1/011001

Preface

In the present volume of *Journal of Physics: Conference Series* we publish the proceedings of the "XXXIX Symposium on Nuclear Physics", that was held from January 5-8, 2016 at the Hacienda Cocoyoc, Morelos, Mexico. The proceedings consist of 20 contributions that were presented as plenary talks at the meeting. The abstracts of all contributions, invited talks and posters, were published in the Conference Handbook.

The Symposium on Nuclear Physics has a long and distinguished history. From the beginning it was intended to be a relatively small meeting designed to bring together some of the leading nuclear scientists in the field. Its most distinctive feature is to provide a forum for specialists in different areas of nuclear physics, both theorists and experimentalists, students, postdocs and senior scientists, in a relaxed and informal environment providing them with a unique opportunity to exchange ideas. After the first meeting in Oaxtepec in 1978, the Symposium was organized every year without interruption which makes the present one the 39th in a row.

The scientific program consisted of 29 invited talks and a poster session on a wide variety of hot topics in contemporary nuclear physics, ranging from the traditional fields of nuclear structure and nuclear reactions to radioactive beams, nuclear astrophysics, hadronic physics, fundamental symmetries and relativistic heavy ions, as well as progress reports of large international projects like the HAWC Observatory in Puebla, Mexico, and the ATLAS and ALICE Collaborations of the LHC accelerator at CERN, Switzerland. In addition, there were several contributions highlighting interesting new results from foreign laboratories like Notre Dame, RIKEN, Jefferson Lab, Oak Ridge, INFN-Legnaro and INFN-LNS, as well as Mexican laboratories at ININ, LEMA and the Carlos Graef Laboratory at IF-UNAM. On the theoretical side there were talks on recent developments in nuclear structure, weakly bound nuclei, cluster models, baryon spectroscopy and non-perturbative QCD.

The high quality of the talks, the prestige of the speakers and the broad spectrum of subjects covered in the meeting, shows that nuclear physics is a very active area at the frontier of scientific research which establishes bridges between many different disciplines.

Luis Acosta
Roelof Bijker

Journal of Physics: Conference Series **730** (2016) 011001 doi:10.1088/1742-6596/730/1/011001

Plenary Talks

- Luis Acosta (IF-UNAM)
  Campaign of measurements to probe the good qualities of the new array FARCOS for multi-fragmentation spectroscopy
- Elí Aguilera (ININ)
  Near-barrier fusion of proton- and neutron-halo systems
- Victoria Isabel Araujo Escalona (IF-UNAM)
  Study of ^{26}Al *production: implications in nucleosynthesis and cosmology*
- Dan Bardayan (Notre Dame)
  Nuclear astrophysics measurements with the TwinSol Radioactive Beam Facility
- Libertad Barrón-Palos (IF-UNAM)
  Parity-violating asymmetries in the interaction of polarized slow neutrons with light nuclei
- Roelof Bijker (ICN-UNAM)
  Baryon spectroscopy in the unquenched quark model
- Francesco Cappuzzello (INFN-LNS)
  The nuclear matrix elements of $0\nu\beta\beta$ *and the NUMEN project at INFN-LNS*
- Manuela Cavallaro (INFN-LNS)
  Signatures of the giant pairing vibration in the ^{14}C *and* ^{15}C *nuclei*
- Efraín Rafael Chávez Lomelí (IF-UNAM)
  IF-UNAM's nuclear physics facilities: new data and extensions
- Christopher Crawford (Kentucky)
  Update on the neutron electric dipole moment
- Raquel Crespo (Lisbon)
  Breakup/knockout of exotic nuclei from collisions with a proton target: developments and perspectives
- César Fernández Ramírez (ICN-UNAM)
  Amplitude analysis of KN scattering
- Moshe Gai (Connecticut)
  The triangular $\mathcal{D}_{3h}$ *symmetry of* ^{12}C
- Alfredo Galindo-Uribarri (ORNL)
  Overview of experimental physics results at Oak Ridge National Laboratory: from neutrinos to clusters
- Adiv González Múñoz (IF-UNAM)
  Latest news from the High Altitude Water Cherenkov Observatory
- Arturo Gómez-Camacho (ININ)
  Different aspects on the effect of ^{6}Li *resonances on elastic scattering*
- Peter Hess (ICN-UNAM)
  The concept of forbiddenness in cluster physics
- Sergio Arturo Iga (ICN-UNAM)
  Color reconnection: a fundamental ingredient of the hadronisation in pp collisions
- Ivano Lombardo (Napels)
  Structure of light nuclei with low energy nuclear reactions
- Arturo Menchaca-Rocha (IF-UNAM)
  Antinuclei and (anti-)hypernuclei studies at the LHC
- Enrique López Moreno (FC-UNAM)
  Phase transitions for rotational states within an algebraic cluster model

- Joel de Jesús Mendoza Temis (ICN-UNAM)
  On the robustness of the r-process in neutron-star mergers against changes of nuclear masses
- Takashi Nakamura (Tokyo)
  Structure of neutron drip-line nuclei
- Stuart Pittel (Delaware)
  Electromagnetic properties of the $I^P = \frac{11}{2}^-$ *isomers in the Cd isotopes: what can we learn from them?*
- Antonio Ortiz (ICN-UNAM)
  Heavy-ion results at the LHC: new discoveries and puzzles
- Oliver Stelzer-Chilton (TRIUMF)
  ATLAS at the LHC: recent results from Run 2
- Adam Szczepaniak (Indiana)
  The Joint Physics Analysis Center
- Víctor Velázquez (TRIUMF)
  Quantum complexity in ^{48}Ca *and* ^{48}Ti
- Tochtli Yépez (La Plata)
  Analysis of a QCD motivated Hamiltonian in the low-energy regime

Peer review statement

All papers published in this volume of *Journal of Physics: Conference Series* have been peer reviewed through processes administered by the proceedings Editors. Reviews were conducted by expert referees to the professional and scientific standards expected of a proceedings journal published by IOP Publishing.

Journal of Physics: Conference Series **730** (2016) 012001 doi:10.1088/1742-6596/730/1/012001

Campaign of measurements to probe the good performance of the new array FARCOS for spectroscopy and correlations.

L. Acosta[1,2,*], R. Andolina[3], L. Auditore[4], C. Boiano[5], G. Cardella[2], A. Castoldi[5,6], M. D'Andrea[2], E. De Filippo[2], S. De Luca[4], D. Dell'Aquila[7], L. Francalanza[7], B. Gnoffo[2], C. Guazzoni[5,6], G. Lanzalone[8,9], I. Lombardo[7], N. Martorana[3,9], T. Minniti[4], S. Norella[4], A. Pagano[2], E.V. Pagano[3,9], M. Papa[2], T. Parsani[5,6], S. Pirrone[2], G. Politi[2,3], F. Porto[3,9], L. Quattrocchi[4], F. Rizzo[3,9], P. Russotto[2], G. Saccà[2], A. Trifirò[4], M. Trimarchi[4], G. Verde[2,10], M. Vigilante[7] and P. Zambon[5,6,11]

1 Instituto de Física, Universidad Nacional Autónoma de México, Cd.Mx., Mexico
2 INFN-Sezione di Catania, Catania, Italy
3 Dip. di Fisica e Astronomia, Università di Catania, Catania, Italy
4 Dip. di Scienze Matematiche e Informatiche, Scienze Fisiche e Scienze della Terra, Università di Messina & INFN Gruppo Collegato di Messina, Messina, Italy
5 INFN Sezione di Milano, Milano, Italy
6 Politecnico di Milano, Dip. di Elettronica, Informazione e Bioingegneria, Milano, Italy
7 INFN-Sezione di Napoli & Dip. di Fisica, Università di Napoli Federico II, Napoli, Italy
8 Università Kore Enna, Enna, Italy
9 INFN Laboratori Nazionali del Sud, Catania, Italy
10 Institute de Physique Nucléaire d'Orsay, Orsay, France
11 Dectris Ltd, Baden, Switzerland

E-mail: *`acosta@fisica.unam.mx`

Abstract.
During the last four years, several measurements have been carried out where the capabilities of FARCOS array were tested. In some of this occasions, FARCOS was coupled to the 4π array CHIMERA, permanently placed at INFN-Laboratori Nazionali del Sud, Catania in order to be tested in real experimental measurements. At the present situation, the FARCOS demonstrator is formed by 4 telescopes out of the originally 20 that will constitute the final array. Here are presented some preliminary results obtained with the new array, probing its qualities and showing the effectiveness of FARCOS telescopes. The initial encouraging results support the construction of the complete array.

1. Introduction

The FARCOS array (Fentoscope Array for Correlation and Spectroscopy) is a new detection system under construction. It is formed by telescopes composed of Double-Sided Silicon Strip Detectors (DSSSD) of 300 and 1500 μm thickness in the first and second stages respectively, and a third stage formed by four CsI(Tl) crystals of 60 mm thickness (see Fig. 1), having 132

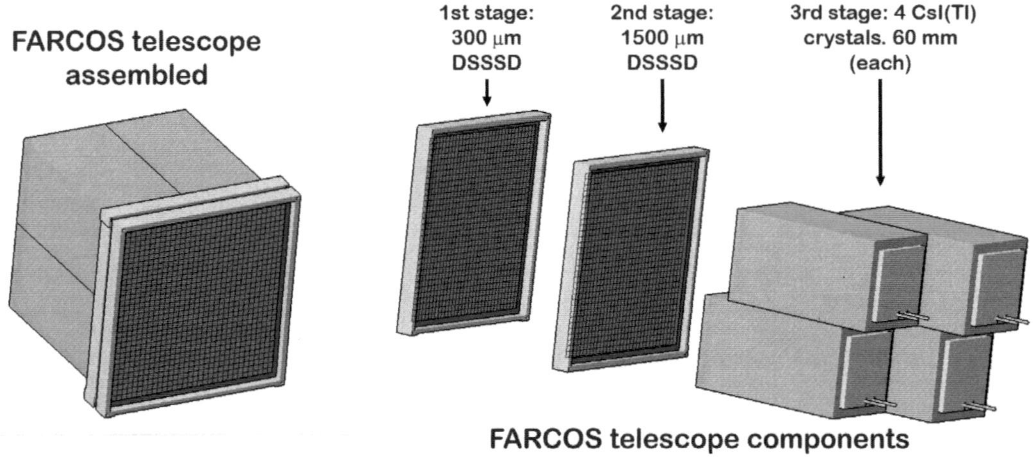

Figure 1. FARCOS telescope composition. Left image shows one of the FARCOS telescope assembled, while in the image on the right are show the components of each telescope: two DSSSD 300 and 1500 μm thickness respectively, and 4 CsI(Tl) crystals (60 mm each) at back, covering the same active area than the silicon detectors (64×64 mm^2).

detection channels in total by telescope. Presently the array (composed of 4 telescopes) is working as demonstrator of the final one, that will be formed by 20 telescopes.

The research program for FARCOS will be to study the particle-particle correlations at intermediate energies and spectroscopy of deep inelastic scattering for stable and radioactive nuclei reactions [1–4]. Moreover, considering the versatility of the array, FARCOS could be used for many other kind of studies.

In particle-particle correlation studies, good reconstruction of the relative energy depends on the angular and energy resolution of the detector. High energy resolution is also required in spectroscopy studies where it is very important to separate close-lying energy states. For this reason, several tests were done to estimate the performance of each identification stage of the FARCOS telescopes, with special attention to energy resolution. During the last 4 years, different array components have been tested, in order to probe the capabilities of the detectors, electronics and digitalization systems [5]. Moreover, FARCOS has been included as part of the set-up in four nuclear physics measurements. At such experiments, FARCOS was used as an auxiliary detector since it was still in the construction phase. In the following sections, some of the preliminary results achieved with the FARCOS demonstrator are shown. The positive results achieved so far, are promising enough to proceed with construction of the full FARCOS array.

2. Qualification of CsI(Tl) crystals.

The first test of the FARCOS components was related with the third stage of each telescope, composed by 4 CsI(Tl) crystals manufactured by Scionix [6] ($60 \times 32 \times 32$ mm^3). The light collected on each crystal is detected by a Hamamatsu 18x18 mm PIN diode S3204-08. In order to test the crystal uniformity and resolution, a test with alpha beam (62A MeV) scattered by ^{208}Pb targets was carried out. Using a DSSSD in front of the crystals, it was possible to map the impact position over the crystal surface, achieving a good method to improve the resolution of the crystal. In Fig. 2 the pixelation of the crystal signal is presented, and how the energy

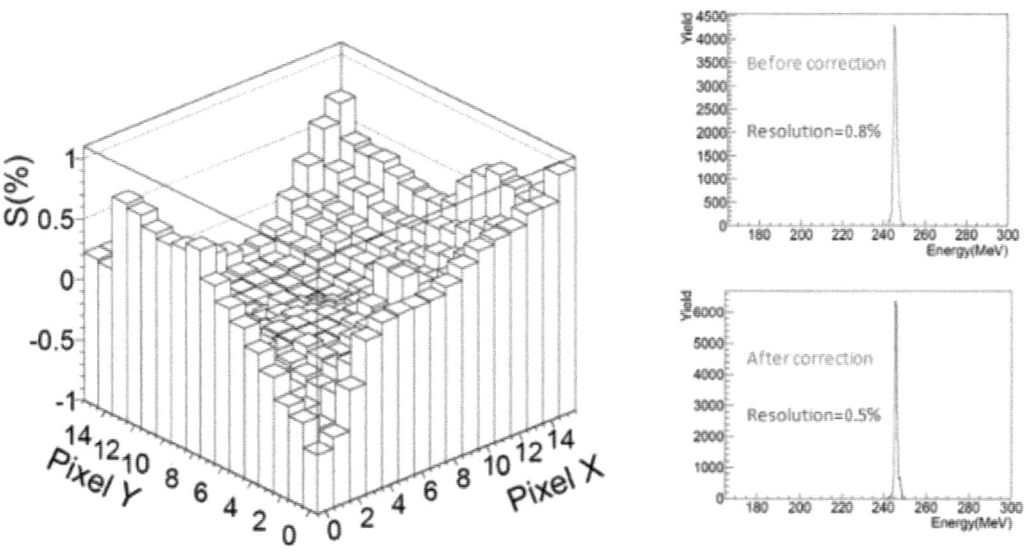

Figure 2. Light response for one CsI(Tl) crystals of FARCOS (left) and energy spectrum before and after correcting for the light response of each pixel with respect to the average response over the entire crystal (right). (Taken from [7]).

signal can be corrected for light response non-uniformities in the crystal. As is shown in Fig. 2 (right side) the energy resolution of CsI(Tl) spectra improve from 0.8% to 0.5%, after taking into account the light response of each pixel with respect to the average response of the crystal. More details of this test of crystals can be found in [7].

In order to test the CsI(Tl) crystals resolution, a standard Charge Sensitive Pre-amplifier (PAC) was used, with gain of 90 mV/MeV to process the signals with two different data acquisition systems: analogue acquisition with commercial 16-channel NIM Amplifiers and a VME QDC and digital acquisition of PAC signals with GET electronics [9]. In both cases the resolution reached was similar, around 3% FWHM, which is sufficient for FARCOS scientific goals. Fig. 3(a) shows, for instance, a triple alpha source spectrum obtained with the standard analogue acquisition system.

3. DSSSD characterization.

For the case of the DSSSD used for FARCOS, several tests have been performed during the last 4 years including resolution tests with analogue and digital electronics and inter-strip effects. The DSSSD's were manufactured by Micron Semiconductor [8], and each detector has 32 vertical strips on the front side and 32 horizontal strips on the rear side, with a total active area of 64 $\times$ 64 mm^2, and resulting in 1024 pixels of 4 mm^2. The signal for each strip is connected to a kapton cable ending with a 34 pin female connector.

For the case of analogue readout, different tests were carried out comparing commercial pre-amplifiers (Mesytec modules [10]) and a dedicated design (32 channels pre-amplifiers developed by INFN-Milano and Politecnico di Milano) [11]. For both configurations amplifiers from CAEN [12], that are currently used for the CHIMERA silicon detectors, were employed. For the analogue set-up, a resolution around 30 keV was achieved for both kinds of pre-amplifiers. Fig. 3(b) shows an example of a triple alpha spectrum for the analogue electronic chain. By using digitizers, the resolution can be improved as was demonstrated when the tests were carried out using SIS3301 [13] digitizing modules at the output of a Anti-Aliasing Amplifiers (AAA) [14].

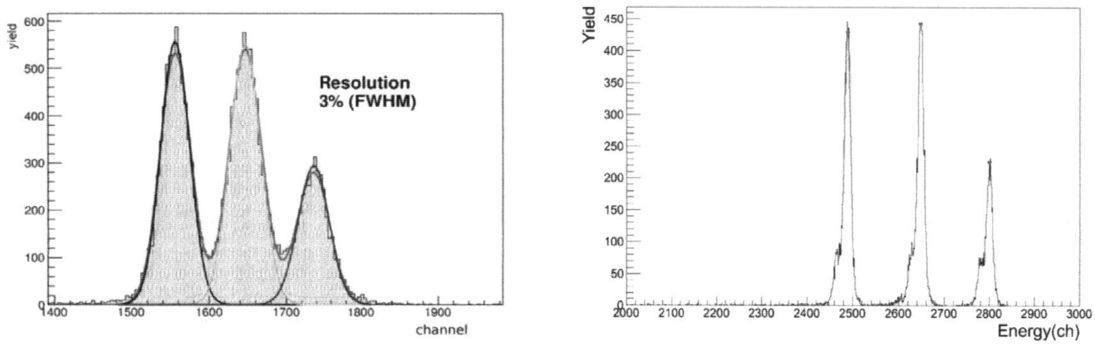

(a) (b)

Figure 3. (a) Triple alpha spectrum for a FARCOS CsI(Tl) crystal by using analogical readout and standard DAQ, achieving a good resolution of 180 keV. (b) Triple alpha spectrum for a FARCOS 1500 μm DSSSD obtained with analogical electronics and standard DAQ. The resolution achieved is around 26 keV.

Using this electronic chain it was possible to reach a resolution around 15 keV for each strip on DSSSD's. The Fig. 4 shows the proton pulse beam calibration spectra taken from the data measured in the DEFEL-line at LABEC facility at Florence, Italy, during experiments dedicated to the study of inter-strip effects. A detailed description of these experiments can be consulted in [15].

4. Testing of FARCOS telescopes with nuclear reactions.

In order to test the capabilities of FARCOS demonstrator, it was mounted in four different measurements as part of the set-up of each experiment. These campaign of experiments was carried out during the last two years, in all of them the telescopes of FARCOS were placed in different geometries, according with the experiment needs. The studies to support each measurement will be described in further publications where a detailed description will be presented. Here we will show just some of the preliminary results achieved with FARCOS. The experi-

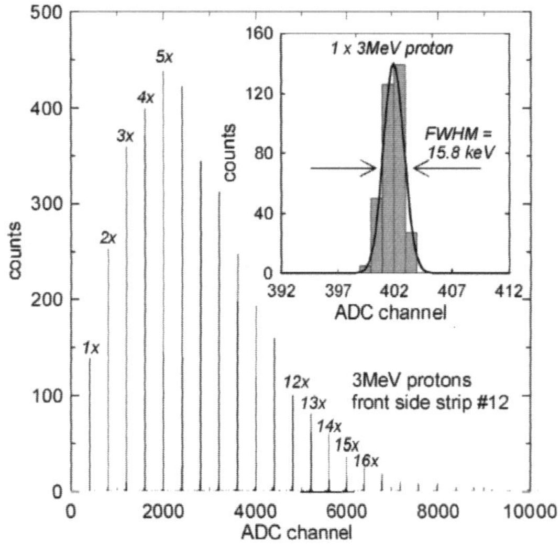

Figure 4. Energy spectrum collected in the case of 3 MeV proton interaction on one strip of the front side with a pulse beam from DEFEL line at LABEC, INFN-Florence. The number of ions in a bunch reflects their statistical distribution in the original beam; as a result, the spectrum envelope is roughly a poissonian distribution. The insets show a detail of the single proton peak together with its Gaussian fitting that allows extracting the peak centroid and FWHM. With this system, the resolution improves 50% with respect to the standard electronic chain (taken from [15]).

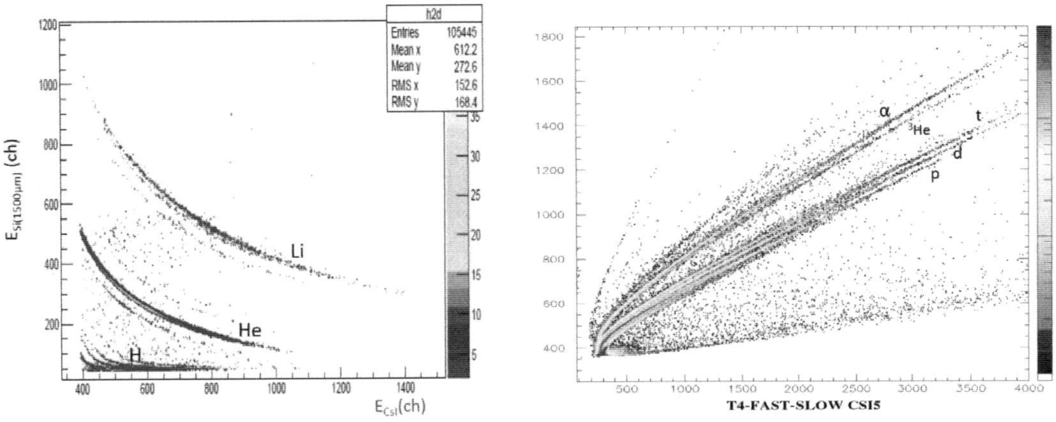

(a) (b)

Figure 5. Preliminary spectra from INKIISSY experiment data related with FARCOS demonstrator. (a) E-ΔE spectrum combining 1500 μm DSSSD (ΔE) and CsI(Tl) (E) signals. (b) Fast-Slow spectrum coming from the particles detected by CsI(Tl) crystals. Both spectra show the excellent resolution of a FARCOS telescope for mass and charge separation.

ments have the following labels: INKIISSY [16–18], CLIR [19], SIKO [20] and PIGMY [21]. All of them were performed at Laboratori Nazionali del Sud, Catania, Italy, employing the CHIMERA multi-detector array [22] as the primary detection system. We present here some of the preliminary results related with FARCOS obtained during two of these measurements: INKIISSY and PIGMY.

During the INKIISSY experiment, many nuclei were produced coming from the ^{124}Xe+^{64}Zn reaction at 35A MeV. In this experiment, the FARCOS demonstrator was placed at $\theta_{Lab} = (20-40)^{\circ}$ from the reaction plane, covering a solid angle where typically 8 CHIMERA telescopes are mounted as part of the spherical detection configuration of the array. The preliminary results for particle identification using the FARCOS components are presented in Fig. 5a and b, where E-ΔE and Fast-Slow spectra are shown. For the particle identification spectrum using 1500 μm DSSSD and CsI(Tl) detectors, it is possible to see with an excellent energy resolution the isotopes corresponding to ^{1,2,3}H, ^{3,4,6}He and ^{6,7}Li; at the same time, using Fast-Slow technique with the CsI(Tl) crystals, p, d, t, ^{3}He and ^{4}He isotopes are as well clearly identified.

For the PIGMY experiment, FARCOS was placed close to $\theta_{Lab} \sim 0^{\circ}$, in order to have information about production of the ^{68}Ni nucleus from the fragmentation generated by the ^{70}Zn+^{9}Be reaction at 40A MeV. The spectrum in Fig. 6 shows the mass and charge separation using the E-ΔE technique. On the spectrum, masses 65, 67, 68, 69 and 71 can be easily identified, making possible the isolation of ^{68}Ni between many other species, which was part of the goals of the experiment.

5. Perpectives and conclusions.

We showed in the previous sections the very good performance of the FARCOS array. The performance assessment done so far open the way to the new development phase, taking place in the next two years.

During this next stage of development, dedicated very-large-scale integration (VLSI) charge

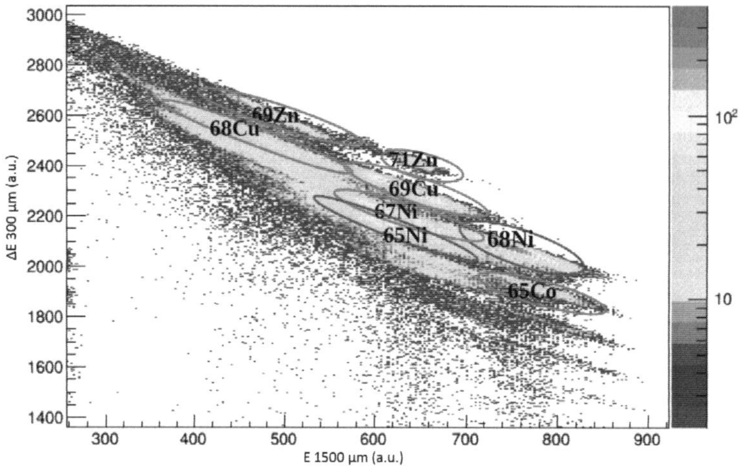

Figure 6. Energy spectra coming from data taken with FARCOS system at the PIGMY experiment. Mass and charge separation (^{68}Ni, for instance) can be easily observed by using the E-ΔE technique in silicon detectors. See the text for more details.

pre-amplifiers will be designed for the DSSSD and CsI(Tl) crystals. Considering the good resolution observed with digital data acquisition, the new electronics for the readout has been particularly designed to be compatible with this kind of system. The improved telescope will equipped with an ASIC (for DSSSD and crystals) coupled to the telescope, in order to make a compact cluster, where just a few cables will control everything coming from the detectors (bias, test, signal and low-voltage). A preliminary design of the new telescope is show in Fig. 7. Moreover, a complete mechanical system is under design, in order to hold the complete array (10 telescopes first, and 20 for the whole array). This new structure will be designed not just to hold a given geometry of the array, both also to stabilize the temperature of the system.

In conclusion, the FARCOS demonstrator has shown very good characteristics in position and energy resolution, even when used as a telescope. A variety of tests of electronic devices have resulted in digital electronics being chosen for the case of Silicon detectors. Several tests that have been carried out with FARCOS demonstrator have proven the very good performance presented by such a system. Even under realistic conditions, the FARCOS array

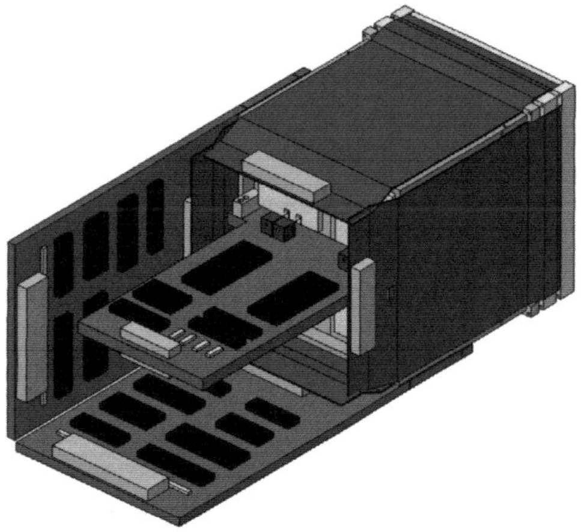

Figure 7. Rear view of the new design for a FARCOS telescope. Four dedicated ASIC boards will be mounted to the telescope (two are shown in the drawing), to manage the signals and bias voltages from DSSSD's, to which are connected by means of kapton cables. In the rear-middle, a single circuit will manage voltages and signals for four CsI(Tl) crystals.

has been a valuable plug-in supporting for the experiment and collecting data relevant for such measurements. During the coming years FARCOS will continue being used for further tests of new electronics and to support other measurements.

Acknowledgments
This work has been partially supported by UNAM-PAPIIT IA101616.

References

[1] Verde G *et al.* 2006 *Eur. Phys. J. A* **30** 81
[2] Acosta L *et al.* 2012 *Eur. Phys. J. Web of Conf.* **31** 00035
[3] Verde G *et al.* 2013 *J. Phys. Conf. Ser.* **420** 012158
[4] Pagano E V *et al.* 2015 *Eur. Phys. J. Web of Conf.* **88** 00013
[5] FARCOS, Technical Design Report. NEWCHIM LNS-INFN 2015
https://drive.google.com/file/d/0B5CgGWz8LpOOc3pGTWdOcDBoWFE/view?pref=2&pli=1
[6] http://www.scionix.nl/
[7] Quattrocchi L *et al.* 2014 *Eur. Phys. J. Web of Conf.* **66** 11001
[8] http://www.micronsemiconductor.co.uk/
[9] Pollacco E *et al.* 2012 *Phys. Procedia* **37** 1799
[10] http://www.mesytec.com/
[11] Acosta L *et al.* 2012 *IEEE Nucl. Sci. Symp. Conf. Rec.* N20-6
[12] http://www.caen.it/
[13] SlSllOOl3100 - User Manual, Struck Innovative System GmbH. Hamburg, Germany, 2008
[14] Boiano C, Guazzoni C, Guazzoni P, Zetta L, and Pagano A 2010 *IEEE Nucl. Sci. Symp. Conf. Rec.* 1389
[15] Acosta L *et al.* 2012 *IEEE Nucl. Sci. Symp. Conf. Rec.* N14-217
[16] Russotto P *et al.* 2014 *J. Phys. Conf. Ser.* **515** 012020
[17] Quattrocchi L *et. al.* 2015 *Eur. Phys. J. Web of Conf.* **88** 01014
[18] De Filippo E *et al.* 2016 *Eur. Phys. J. Web of Conf.* in press
[19] Dell'Aquila D *et al.* 2016 *Phys. Rev. C* **93** 024611
[20] Verde G and Kokalova T 2012 PAC-LNS INFN
[21] Cardella G *et al.* 2013 PAC-LNS INFN
[22] Pagano A *et al.* 2004 *Nucl. Phys. A* **734** 504

Journal of Physics: Conference Series **730** (2016) 012002 doi:10.1088/1742-6596/730/1/012002

Near-barrier fusion of proton- and neutron-halo systems

E F Aguilera

Departamento de Aceleradores, Instituto Nacional de Investigaciones Nucleares, Apartado Postal 18-1027, C. P. 11801, México, D. F., México

E-mail: eli.aguilera@inin.gob.mx

Abstract. It is shown that the behaviour of the fusion excitation functions for proton-halo and neutron-halo systems presents important differences, especially in the energy region slightly above the barrier. Measurements for ^{6}He, ^{11}Li and ^{11}Be projectiles are discussed to exemplify the behaviour of neutron-halo systems, while experiments with ^{8}B beams illustrate the situation for proton-halo nuclei. With respect to a standard benchmark, neutron- (proton-) halo systems show a fusion suppression (enhancement) above the barrier.

1. Introduction

In the later two decades, fusion cross sections near the Coulomb barrier have been measured for several neutron-halo systems, but only recently a couple of true proton-halo systems (*i.e.*, having a proton-halo in its g.s.) have been studied in this context.

The aim of this work is to show that the behaviour of the fusion excitation functions for halo systems presents important differences depending on whether the halo is charged or neutral. More precisely, it will be shown that proton-halo and neutron-halo systems behave differently, especially in the energy region slightly above the barrier.

The work starts with a reminder about the behavior of neutron-halo systems, whose main feature is that they present a fusion suppression above the barrier (Sec. 2). Then, in Sec. 3 some apparently controversial data will be shown, corresponding to two proton-halo systems: ^{8}B + ^{58}Ni (Sec. 3.1) and ^{8}B + ^{28}Si (Sec. 3.2). These data have been critically reviewed recently, as described below. On one hand, a possible contamination in the B on Ni data was carefully analyzed. In addition, since in each case a mapping of the measured observable into the fusion cross section had to be made by using statistical model calculations, it became necessary to consider possible model dependencies. As a conclusion, it will be shown in Section 3.3 that the data for both systems are actually consistent with each other and with a fusion enhancement above the barrier. Finally, a summary and the main conclusions of this work will be presented in Section 4.

2. Neutron-halo systems

It's been long known that neutron-halo systems present a fusion suppression above the barrier. By adopting the parameter-free São Paulo Potential (SPP) [1] as an appropriate bare potential to describe fusion reactions, Crema *et al.* [2] showed that the experimental fusion excitation functions for the neutron-halo projectile ^{6}He on targets of ^{64}Zn, ^{209}Bi and ^{238}U present very similar behavior. They found that, for energies above the respective Coulomb barriers, the experimental fusion cross sections for all three systems were suppressed with respect to the corresponding SPP predictions. In an earlier work [3], the same authors had shown that the SPP gives reliable results in the study of fusion reactions with stable weakly-bound projectiles. A realistic density for ^{6}He was used in Ref. [2] for the calculation of the respective potentials.

In order to systematize this kind of results, it is convenient to use an appropriate method of data reduction that allows one to compare with each other, in the same plot, the cross sections for different systems. One method that has been successfuly used to compare fusion data consists of using the following prescription to reduce the cross section and the energy [4, 5, 6, 7]:

$$\sigma_{Red} = \frac{2E}{\hbar\omega {R_b}^2}\sigma, \qquad E_{Red} = \frac{E - V_b}{\hbar\omega}, \tag{1}$$

where $V_b, R_b, \hbar\omega$ are the corresponding barrier parameters. Extracting these parameters from a realistic bare potential for each system, such as the SPP, provides a common benchmark for a meaningful comparison of different systems [5, 6, 7]. The above prescription is based on a well-known analytic expression for the cross section, σ^W(E), which was derived long ago by Wong [8]. If applied to σ^W(E), one gets the so called universal fusion function (UFF) [6, 7],

$$\sigma^W_{Red}(E_{Red}) = \frac{2E}{\hbar\omega {R_b}^2}\sigma^W(E) = ln[1 + e^{(2\pi E_{Red})}]. \tag{2}$$

A comparison of fusion data for the neutron-halo systems ^{6}He + ^{209}Bi [9] and ^{6}He + ^{238}U [10] was done in Refs. [6, 7] using this prescription. For both systems, the authors found a suppression factor of 0.7 with respect to the UFF. This is illustrated in Fig. 1, where in addition we have included data for ^{6}He + ^{64}Zn [11], ^{11}Li + ^{208}Pb [12], and ^{11}Be + ^{209}Bi [13]. One can see that, for all these neutron-halo systems, fusion is suppressed for energies above the barrier, with suppression factors that are between ~ 0.5-0.7.

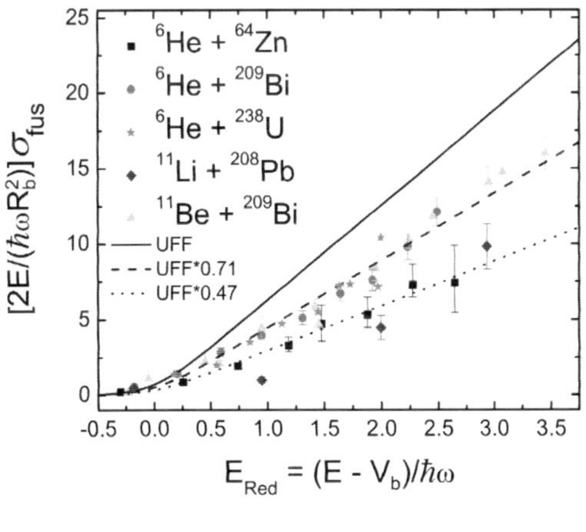

Figure 1. Reduced fusion cross sections for several neutron-halo systems. Data from: ^{6}He + ^{209}Bi [9], ^{6}He + ^{238}U [10], ^{6}He + ^{64}Zn [11], ^{11}Li + ^{208}Pb [12], and ^{11}Be + ^{209}Bi [13].

The barrier parameters that were used to make the reduced plot of Fig. 1 are given in Table 1. They were obtained from the double-folding SPP, using default values for the respective matter and charge densities, which follow the systematics observed for many nuclei. The observed deviations from the UFF in Fig. 1 can thus be ascribed either to static effects related to deviations in the actual densities, or to dynamic effects, originating from some intrinsic properties of the respective nuclei.

Journal of Physics: Conference Series **730** (2016) 012002 doi:10.1088/1742-6596/730/1/012002

Table 1. Barrier parameters used to reduce the data for all systems considered in the present work, obtained in each case from the respective São Paulo Potential.

System	Reference	R_B (fm)	V_B (MeV)	$\hbar\omega$ (MeV)
^{6}He + ^{64}Zn	[11]	9.5	8.4	2.9
^{6}He + ^{209}Bi	[9]	11.6	19.3	3.9
^{6}He + ^{238}U	[10]	11.9	27.8	4.0
^{11}Li + ^{208}Pb	[12]	12.0	31.66	3.44
^{11}Be + ^{209}Bi	[13]	11.8	38.2	4.0
^{8}B + ^{28}Si	[20]*	8.18	11.27	3.38
^{8}B + ^{58}Ni	[14]*	8.9	20.8	4.09

* reanalyzed in Ref. [17]

3. Proton-halo systems

Proton-halo systems have been far more elusive. Only two such systems have been measured, both corresponding to ^{8}B projectiles but with different targets. It is generally agreed that the ^{8}B nucleus has a one-proton halo configuration in its ground state. An apparent controversy aroused between the two datasets for these proton-halo systems, but a recent work showed that they are actually consistent with each other. This will be briefly described here.

3.1. ^8B + ^{58}Ni

Experimental cross sections for the fusion of ^{8}B + ^{58}Ni were published in 2011 [14]. With respect to expectations of a simple barrier penetration model calculation, they presented a big enhancement, even for energies above the Coulomb barrier, which in this case is at 20.8 MeV. This differs from what has been observed for neutron-halo systems, as discussed in Sect. 2 above. These results were also surprising because the behaviour of total reaction cross sections for proton-halo and neutron-halo systems is quite similar. Indeed, it has been shown that the reduced total reaction cross sections for ^{8}B + ^{58}Ni and for several neutron-halo systems fall on the same trajectory when plotted as a function of the reduced energy [15].

In Ref. [14], evaporation protons were measured at backward angles and the statistical model code PACE2 [16] was used to calculate the respective multiplicity, which was then used to deduce the corresponding fusion cross section. An advantage of this technique is that the respective proton multiplicities are fairly high, which enhances the probability of observing the fusion events. In addition, model calculations are expected to be more stable for high multiplicities. A possible disadvantage of the technique is related to the fact that protons are also produced in the breakup (bu) of ^{8}B into ^{7}Be + p. If these bu protons reach the detectors, there is no way to distinguish them from evaporation protons.

The contribution of bu protons to the measured proton cross sections, σ_p, was estimated in Ref. [17] by using CDCC calculations available in the literature. Except for the lowest energy measured, this contribution was practically insignificant and, furthermore, the lowest energy point could be recovered by analyzing the evaporation alpha particles that were detected during the same experiment. In fact, alpha particle measurements were also liable to be analyzed for a few more energies and they gave fusion cross sections consistent with those obtained from the proton measurements at the same energies [17].

In addition, possible model dependencies in the $\sigma_p \rightarrow \sigma_{fus}$ mapping were extensively studied in Ref. [17] by using three different evaporation codes (PACE2 [16], LILITA [18], and CASCADE [19]) and carefully choosing the respective input parameters. The relevant proton multiplicities, M_p, were calculated using two different criteria. First, the three evaporation codes were run using always the same set of optical-potential parameters (OMP) to calculate the transmission coefficients corresponding to neutrons, protons and alpha particles. The results are displayed with solid squares, circles, and up-triangles in Fig. 2. From the average of these results, a proton multiplicity $< M_p > \pm\ \delta < M_p >$ was obtained for each energy (down-triangles in Fig. 2) and corresponding fusion cross sections $\sigma_{fus} = \sigma_p/< M_p >$ were calculated. Then, a different set of OMP's was used in CASCADE, corresponding to those recommended by the author of this code, which gave proton multiplicities M'_p (stars in Fig. 2). These latter multiplicities happened to be very similar to $< M_p >$, which indicates a fairly low sensitivity of proton multiplicities to the OMP's used in the calculations for this particular system. In other words, the effects of model dependency in the $\sigma_p \rightarrow \sigma_{fus}$ mapping are rather weak and can be reasonably accounted for by the uncertainties $\delta < M_p >$ assigned to $< M_p >$. As a conclusion, the ^{8}B + ^{58}Ni fusion cross sections reported in Ref. [14] did not change significantly. In addition, the respective systematic error originally reported was eliminated and replaced by a proper folding of $\delta < M_p >$ into the global uncertainty.

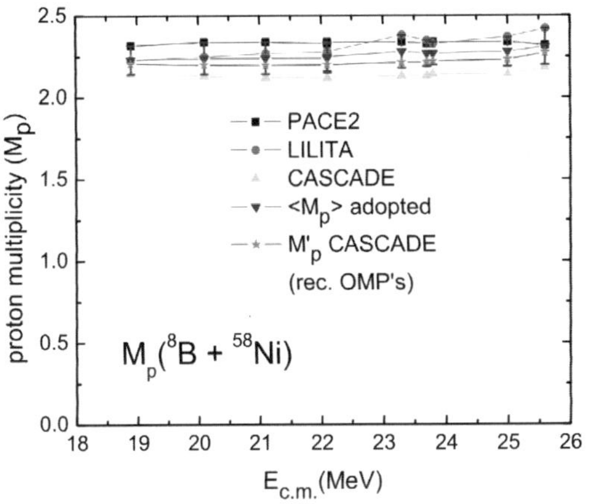

Figure 2. Proton multiplicities obtained for the ^{8}B + ^{58}Ni system with the three codes PACE2, LILITA, and CASCADE, as described in the text.

3.2. $^8B + {}^{28}Si$

In the experiment with the Si target [20], alpha particles were measured instead of protons. The detector covered an angular range at forward angles, but no angular distribution was actually measured. The alpha multiplicity used to deduce the fusion cross section was calculated in this case with the code CASCADE [19]. A good asset in this technique is the fact that alpha particles coming from direct processes are unlikely, but the low alpha multiplicities and the necessary dependency on the detector efficiency become important disadvantages in this technique. In contrast to the results for ^{8}B + ^{58}Ni, the fusion cross sections reported in Ref. [20] for ^{8}B + ^{28}Si showed a supression above the barrier.

As for the detector efficiency, unfortunately this was not discussed in Ref. [20], but the respective uncertainty would probably translate into an additional systematic error larger than 15% in the corresponding data, as discussed in Ref. [17]. In addition, for this system an equivalent analysis using the same three codes (PACE2, LILITA, CASCADE) also was carried out in Ref. [17]. A higher degree of model dependency was found for the respective $\sigma_\alpha \rightarrow \sigma_{fus}$ mapping, which is most probably a consequence of the low alpha multiplicities M_α. This is illustrated in Fig. 3, where the symbols have an eqivalent meaning to those of Fig. 2, but M_p is replaced by M_α and ^{58}Ni is replaced by ^{28}Si. The same criteria described in Sec. 3.1 to obtain $< M_p >$ and M'_p were used here to obtain $< M_\alpha >$ and M'_α, respectively. In this case, the values of M'_α fell out of the uncertainties assigned to $< M_\alpha >$, indicating a fairly strong model dependency. This is the reason why the respective results for σ_{fus} were plotted in Ref. [17] as two separate datasets, one corresponding to $< M_\alpha >$ and the other corresponding to M'_α. In next Section, a proper average of these two datasets for ^{8}B + ^{28}Si will be plotted instead.

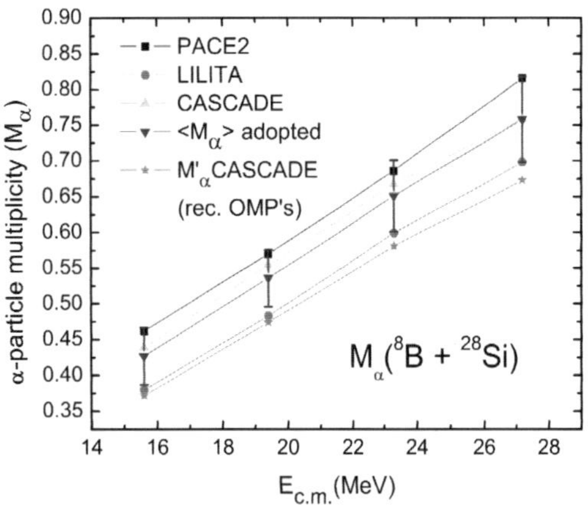

Figure 3. α-particle multiplicities obtained for the ^{8}B + ^{28}Si system with the three codes PACE2, LILITA, and CASCADE, as described in the text.

3.3. Consistency of fusion data for ^8B + (^{28}Si, ^{58}Ni)

Figure 4 shows a reduced plot of fusion data for the ^{8}B + (^{28}Si, ^{58}Ni) systems, as obtained from the reanalysis of Ref. [17] that was described in Sections 3.1 and 3.2. The barrier parameters used to reduce the data are given in Table 1. The data points for the ^{28}Si target were actually taken as the average of the values obtained under the two model assumptions which led to multiplicities $< M_\alpha >$ and M'_α, as described above. Further details of the respective statistical model calculations can be found in Ref. [17].

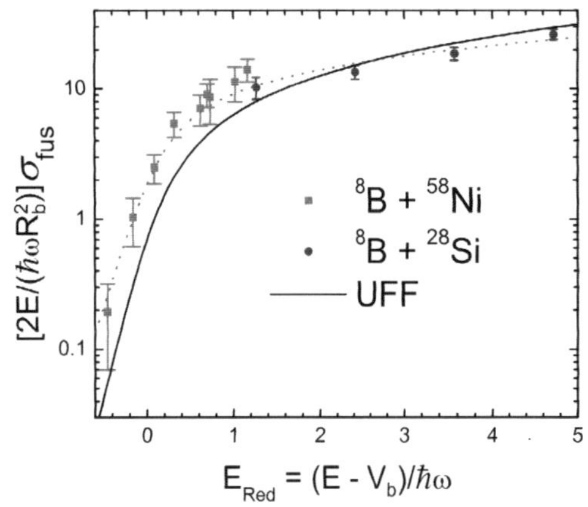

Figure 4. Reduced fusion cross sections for the proton-halo systems ^{8}B + (^{58}Ni, ^{28}Si). Datasets for both systems are from the reanalysis made in Ref. [17] of the original data [14, 20]. The dotted line is to guide the eye.

It can be seen that the reduced fusion cross sections for the two proton-halo systems are certainly consistent with each other when the effects of model dependency are properly taken into account. It is worth mentioning also that a possible systematic error larger than 15%, associated to the respective detector efficiency, should probably be assigned to the ^{8}B + ^{28}Si data. The mentioned controversy between the two proton-halo systems was thus solved. The dotted line in Fig. 4, which was drawn to guide the eye, clearly shows that both datasets are also consistent with a fusion enhancement above the barrier. For $E_{Red} = 1$, this curve indicates an enhancement factor of 1.5. This reduced energy corresponds to 4.1 MeV (3.4 MeV) above the barrier for the case of the ^{58}Ni (^{28}Si) target.

4. Conclusions

The behaviour of the fusion excitation functions for halo systems is reviewed. Neutron-halo systems present a fusion supression for energies above the barrier, which is illustrated in a reduced plot of experimental data for several systems corresponding to projectiles of ^{6}He, ^{11}Li and ^{11}Be. As for proton-halo systems, an apparent discrepancy in the behaviour of the fusion excitation functions for ^{8}B + (^{28}Si, ^{58}Ni) is shown to disappear when the effects of model dependency are taken into account. Both systems are thus consistent with a fusion enhancement for energies above the barrier. An important difference is thus confirmed in the fusion of halo systems, which depends on whether the halo is charged or neutral.

Acknowledgments

This work has been partially supported by CONACYT (México).

References

[1] Chamon L C, Carlson B V, Gasques L R, Pereira D, De Conti C, Alvarez M A G, Hussein M S, Cândido Ribeiro M A, Rossi Jr. E S and Silva C P 2002 *Phys. Rev. C* **66** 014610
[2] Crema E, Gomes P R S, and Chamon L C 2007 *Phys. Rev. C* **75** 037601
[3] Crema E, Chamon L C, and Gomes P R S 2005 *Phys. Rev. C* **72** 034610
[4] Prasad N V S V, Vinodkumar A M, Sinha A K, Varier K M, Sastry D L, Madhavan N, Sugathan P, Kataria D O, and Das J J 1996 *Nucl. Phys. A* **603** 176
[5] Gasques L R, Chamon L C, Pereira D, Alvarez M A G, Rossi Jr. E S, Silva C P, and Carlson B V 2004 *Phys. Rev. C* **69** 034603
[6] Canto L F, Gomes P R S, Lubian J, Chamon L C, and Crema E 2009 *Nucl. Phys. A* **821** 51
[7] Canto L F, Gomes P R S, Lubian J, Chamon L C, and Crema E 2009 *J. Phys. G: Nucl. Part. Phys.* **36** 015109
[8] Wong C Y 1973 *Phys. Rev. Lett.* **31** 766
[9] Kolata J J *et al.* 1998 *Phys. Rev. Lett.* **81** 4580
[10] Raabe R *et al.* 2004 *Nature* **431** 823
[11] Scuderi V *et al.* 2011 *Phys. Rev. C* **84** 064604
[12] Vinodkumar A M *et al.* 2013 *Phys. Rev. C* **87** 044603
[13] Hinde D J and Dasgupta M 2010 *Phys. Rev. C* **81** 064611
[14] Aguilera E F *et al.* 2011 *Phys. Rev. Lett.* **107** 092701
[15] Kolata J J and Aguilera E F 2009 *Phys. Rev. C* **79** 027603
[16] Gavron A 1980 *Phys. Rev. C* **21** 230
[17] Aguilera E F *et al.* 2016 *Phys. Rev. C* **93** 034613
[18] Gomez del Campo J and Stokstad R G 1981 ORNL Report TM-7295
[19] Pühlhofer F 1977 *Nucl. Phys. A* **280** 267
[20] Pakou A *et al.* 2013 *Phys. Rev. C* **87** 014619; *Phys. Rev. C* **87** 049901(E)

Journal of Physics: Conference Series **730** (2016) 012003 doi:10.1088/1742-6596/730/1/012003

The $^{28}Si(d,\alpha)$ reaction

V Araujo-Escalona[1, a)], L Acosta[1], E Andrade[1], L Barrón-Palos[1], C Cantó[1], F Favela[1], A Huerta[1], M E Ortiz[1], C Solís[1] and E Chávez[1, b)]

[1]Instituto de Física
Universidad Nacional Autónoma de México
Circuito de la Investigación Científica, Ciudad Universitaria
CP. 04510, Coyoacán, Ciudad de México.

[a)] vickyaraujo@fisica.unam.mx
[b)] chavez@fisica.unam.mx

Abstract. Recent techniques and experimental methods open up new ways to explore nuclear reactions relevant to nucleosynthesis at stellar combustion temperatures. In this work we focus on the case of ^{26}Al, isotope associated with the identification of active nucleosynthesis spots in the cosmos. Its presence in the solar system was unexpected until it was found in the Allende meteorite. It is now understood that cosmic rays induce nuclear reactions to produce ^{26}Al. On Earth, this process is well known and is the basis of many environmental studies. So ^{26}Al is not just the product of some high-metallicity star collapse. Natural Silicon targets were irradiated with deuteron beams of energies between 0.9 and 2.6 MeV. The produced ^{26}Al were counted later-on in an AMS facility. In this paper, we describe in detail the experimental protocol developed to quantify the number of deuterons hitting the target, as a first step to determine the total cross section.

1. Introduction

^{26}Al is a remarkable isotope from so many points of view that it is worth reminding a few. From the nuclear structure point of view, it is well known that the heaviest odd-odd stable nucleus is ^{14}N. From there on: ^{18}F, ^{22}Na, ^{26}Al, and up to heavier nuclei, they all decay by β+ emission. The ground state quantum numbers of these three consecutive listed nuclei show a curious sequence: 1+, 3+ and 5+ [1]. The unusual large spin of the ^{26}Al ground state makes the β+ transition to ^{26}Mg ground state (0+) so difficult (fourth forbidden) that it decays instead to the first excited state (1.80873 MeV, 2+), still second forbidden [2], as a result, ^{26}Al has the largest life time (7.2×10^5 y) among all odd-odd N=Z nuclei.

Moreover, the spin assignment of its five lower lying states gives the sequence: 5+, 0+, 3+, 1+ and 2+. Now Electromagnetic transitions (gamma) between these states also involve large angular momentum transfers, making all these states uncharacteristically long lived. So much so, that the first excited state 0+ at 228.305 keV has a mean life of 6.3452 s [1], which for an electromagnetic transition is extremely large. So much so that it is considered a "meta-stable" state and often referred to as ^{26m}Al which plays an important role in studies of the weak interaction (Super-allowed $0^+ \rightarrow 0^+$ transitions) and of the Cabibbo–Kobayashi–Maskawa (CKM) matrix [3].

^{26}Al is like a laboratory for multidisciplinary studies. Besides the ^{14}C/^{12}C ratio, the ^{26}Al/^{27}Al precise determination has become one of the main activities in Accelerator Mass Spectrometry (AMS) laboratories [4].

There is the still the outstanding question of its formation: how ^{26}Al is made? There are many ways in which a nucleus can be formed: Stellar Nucleosynthesis, fragmentation of heavier nuclei by high-energy cosmic particles, neutron capture and low energy nuclear reactions. Cross section measurements from nuclear physics experiments are needed to better understand all these processes.

As for cross-disciplinary sciences, the fact that this isotope is found on Earth, the solar system, the galaxy and the Universe [5], makes of it a unique tool for discovery in geology, environmental sciences, solar models, star and galactic evolution, cosmology and others.

Special interest in ^{26}Al surged with the observation of a ^{26}Mg excess of in the Allende meteorite [6] and grew with the detection of the 1.809 MeV γ-line in cosmic-rays [7]. This γ-ray is a clear indicator of the presence of ^{26}Al β-decay and allows the validation of some stellar nucleosynthesis models. According to these models, ^{26}Al is mainly produced through the ^{25}Mg(p,γ) reaction [8]–[11]. Another reaction that has been studied in this regard is ^{24}Mg(^{3}He,p)^{26}Al [12]. No information can be found in the literature about the contribution from the ^{28}Si(d,α)^{26}Al. This work is a progress report on the study of this last reaction. Deuteron irradiations are made at the 5.5 MV CN-Van de Graaff accelerator facility. The irradiated Silicon targets are taken to our Accelerator Mass Spectrometry (AMS) facility for ^{26}Al counting. The lowest concentration (^{26}Al/^{27}Al ratio) measurable at LEMA being 10^{-14} projects the lowest cross section we could measure with this technique to be around 1 nbarn.

2. Experimental setup

Our protocol is similar to that proposed by Arazi et al. [10], [13] in their study of the ^{25}Mg(p,γ)^{26}Al reaction, where Magnesium targets were first irradiated with proton beams, and after chemical separation and addition of ^{27}Al (pilot), a cathode is prepared to count the ^{26}Al isotopes in an AMS facility. In our case, no chemical preparation of the irradiated substance nor additional pilot material were necessary. The cathode itself as needed for AMS measurements was filled with a known mixture of ^{nat}Si and ^{27}Al powders and directly irradiated with deuteron beams at the 5.5 MV CN-Van de Graaff Accelerator facility. Cathodes are then taken "as is" to LEMA ("Laboratorio de Espectrometría de Masas con Aceleradores") for ^{26}Al counting.

2.1. Target preparation

High purity Silicon and Aluminum powders were mixed and tried as cathodes at the AMS isotope separator in different proportions to select the target composition that would have the most Silicon atoms and yet produce enough ^{27}Al current in the Ion Source. This "pilot" beam is needed to fine-tune the Isotope separator for mass 27 and then the necessary fields for mass 26 are easy to set. At the end, both ^{26,27}Al isotopes have to be counted precisely. The (^{26}Al/^{27}Al) ratio measured is finally compared to that of a known reference material for an absolute value [14].

In [14] is detailed the use of the copper cathode holder required for our AMS measurement. These are the same sample holders used as containers for the Al:Si mixture during deuteron irradiation. Avoiding in this way any additional chemical treatments between deuteron irradiation and AMS counting.

2.2. Deuteron Irradiation

After selecting the Si/Al ratio, fresh cathodes were prepared and taken to the 5.5 MV Van de Graaff accelerator laboratory to be irradiated with deuteron beams.

Two 300 μm thick *PIPS* (Passivated Implanted Planar Silicon) [15], [16], detectors were placed at 150° relative to the beam direction left and right to detect back-scattered particles, arrangement known as Rutherford Backscattering Spectrometry (RBS). Both detectors used 1.8 mm diameter collimators [14]. Since there is no significant difference between the data from both detectors, in this report only data from one detector will be shown for simplicity. At these low energies, far below the Coulomb barrier for this system (3.91 MeV), reaction yields are extremely small and strongly energy dependent.

3. Data analysis and preliminary results

Figure 1 shows the energy spectrum of particles reaching one of our detectors at 150° during an irradiation at E_d = 2.2 MeV. The well-known elastic scattering contribution from an infinite thick target has been simulated using SIMNRA [17], a fully graphical user interface program that can simulate and fit elastic scattering and nuclear reaction data. A higher energy yield from positive Q-value (d,α) and (d,p) reactions is observed. At some point, it was suggested the idea to use this high-energy yield for normalization, however, as the beam energy is reduced this yield disappears, as it should, and since our final goal is to extend our cross section measurements at energies as low as possible, we decided to use the elastic yield.

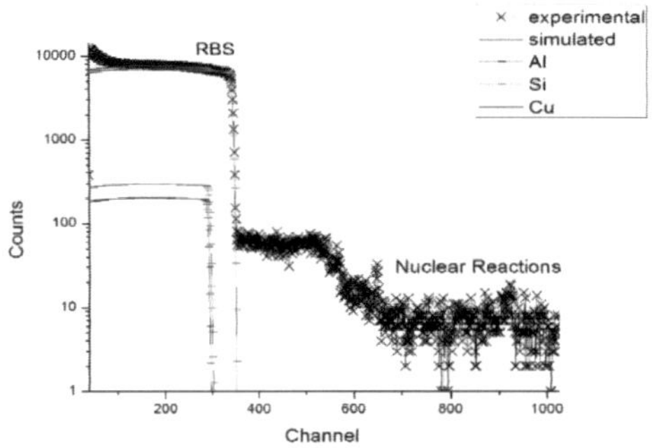

Figure 1. Energy spectrum and simulation by SIMNRA from the deuteron irradiation of the target at E_d = 2.2 MeV

In figure 2 we zoom-in the low energy section (elastic) of two spectra corresponding to deuteron beam energies of 1.7 (a) and 2.2 (b) MeV. Since our energy resolution does not allow the separation from the contribution of mass 27 to that of mass 28, the Al/Si target mixture contribution is taken together as one (mass 28) for the SIMNRA simulations. These simulations allow us to identify the contributions to each spectrum from deuterons scattered off Copper and those off Al/Si. The quality of the fit is clearly not the same for both energies; this fact led us to conclude that there was a misalignment between the beam axis and the center of the target. This misalignment was extensively addressed in a previous report [18]. There we showed how critical beam optics and alignment are when you deal with a 650 μm radius.

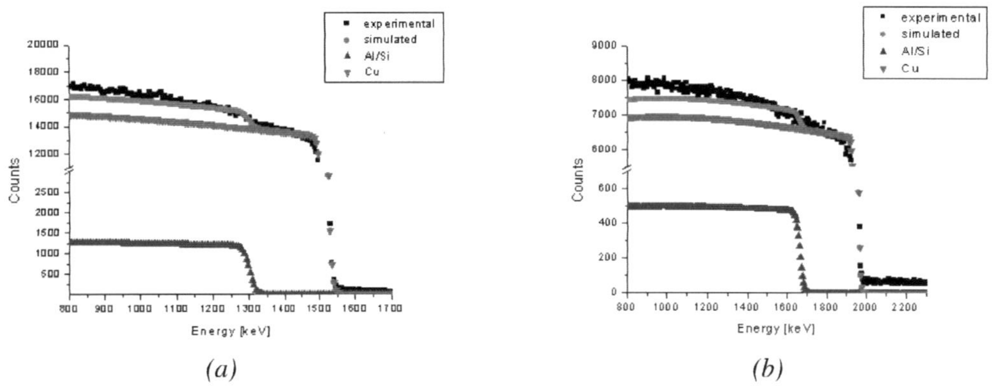

Figure 2. Data simulation from RBS spectra from deuteron irradiation of two targets: E_d= 1.7 (a) and 2.2 (b) MeV bombarding energy.

Journal of Physics: Conference Series **730** (2016) 012003 doi:10.1088/1742-6596/730/1/012003

Determination of the total number of incident deuterons is made by simulating the RBS spectrum with SIMNRA.

The RBS spectra (see Figure 3) show that there is a significant interaction of the beam with the copper holder. The first part of the analysis focuses on the fit of the RBS contribution from Copper. Once a good fit is reached, we subtract this contribution to the original data. The result is considered the contribution to the RBS spectrum from deuteron scattered from the Al/Si mixture. A second fit to this subtracted spectrum with SIMNRA yields the desired number of incoming deuterons during irradiation.

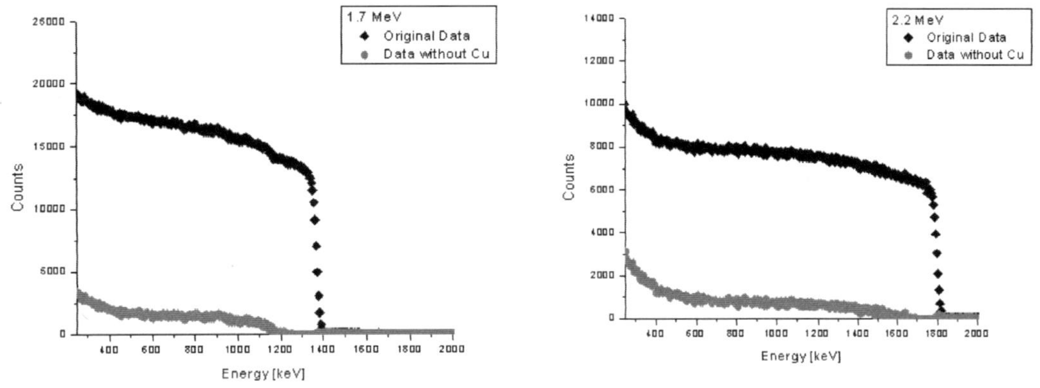

(a) (b)

Figure 3. RBS spectra for two different irradiation energies (a) 1.7 MeV and (b) 2.2 MeV. Squares represent the original data and circles represent only the interaction of the deuteron beam with Si/Al.

3.1. ^{26}Al counting

Irradiated and non-irradiated cathodes were taken to *LEMA* for Aluminum counting. Figure 4 shows typical E-ΔE maps from the split-anode of the ionization chamber detector in the Isotope Separator of *LEMA*. Counts in the shaded region correspond to ^{26}Al events. The pilot beam (^{27}Al) is stopped and quantified in a Faraday Cup. These measurements are compared with those from a reference material to obtain an absolute determination of the excess ^{26}Al produced.

In some of our measurements, data deviated significantly from the "typical" map, as shown in Figure 4 (a). The large unidentified background could be eliminated by using 5 mm collimators all along the isotope separator in LEMA, both for tuning and data taking. Reducing the count-rate, but making sure to exclude unexpected contributions, see Figure 4 (b).

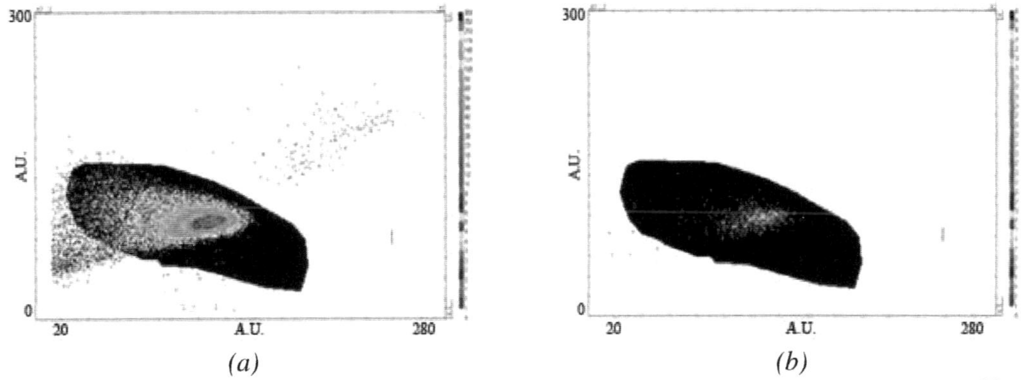

(a) (b)

Figure 4. E-ΔE maps from the ionization chamber at LEMA. The dark region corresponds to ^{26}Al ions (a) Maximum acceptance. (b) 5 mm diameter collimation throughout the system, limiting detection of just the central trajectories.

It is worth stressing at this point that the compound normally used for AMS work with Aluminum, is Aluminum oxide (Al_2O_3) and that we are using a simple mixture of Al/Si.

To better understand the origin of our background, we did a mass analysis of our beam at the injector magnet by using a beam current integrator and slowly changing the magnetic field, see Figure 5. The arrow shows the magnetic field corresponding to mass 27. To the right, at higher magnetic fields, the very large contribution of mas 28 is visible. One can see that mass 27 is "mounted" on some continuous background when the collimators are out, making tuning difficult and eventually leading to inaccurate measurements of the $^{26}Al/^{27}Al$ ratio (we will call it ΔAl in the following).

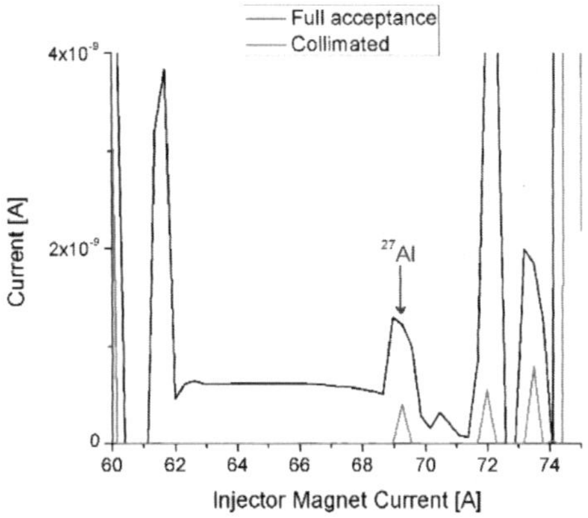

Figure 5. Beam current integration before injecting into the tandem accelerator by scanning the magnetic field in the low energy magnet. The black line corresponds to full acceptance and the red line to the case where only central trajectories are allowed.

If we restrict the beam before the injecting magnet allowing only central trajectories, we get the mass distribution shown in Figure 5 (red line). The ^{27}Al current is much smaller, but the background around mass 27 disappears, the tuning is more certain and we can guarantee that each and every count in the shaded area is ^{26}Al as shown before in Figure 4 (b).

All deuteron-induced nuclear reactions occur at the surface of our Al/Si mixture target. At our energies, beams are stopped within the first 30 μm. Figure 6 shows ΔAl for cathodes irradiated at different beam energies as a function of time for the first 5 minutes. ΔAl exhibits the same behavior for all but one irradiated cathode. The contrast with a control non-irradiated cathode serves to demonstrate that an excess ^{26}Al was indeed produced during deuteron irradiations.

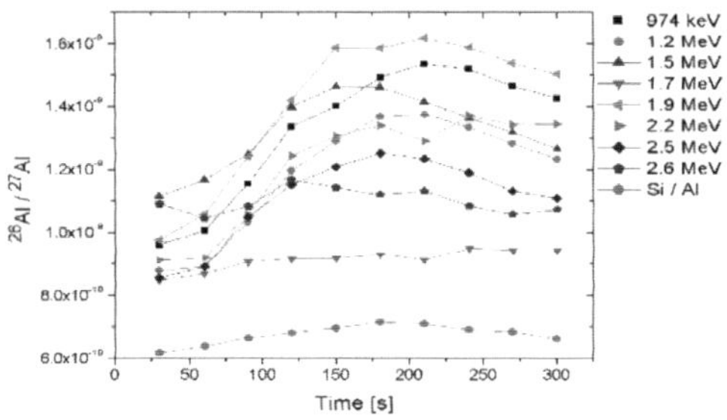

Figure 6. Shows $^{26}Al/^{27}Al$ ratio of cathodes irradiated and compared with a non-irradiated cathode. (Non-standard data)

The Sputtering Negative Ion Cesium Source (SNICS) at *LEMA* will sputter ions from the very center of the cathode. If, as was the case of the irradiation at 1.7 MeV, the beam-target alignment is off by a few hundred microns, the reactions of interest will happen near the edge of the Copper holder and the *SNICS* will not sputter those atoms.

Knowing we were facing systematic problems in our data, we decided to continue with the data processing to calculate a "yield" (thick target integrated cross section), which is one step closer to the desired excitation function.

The number of ^{26}Al is derived from ΔAl in *LEMA* and the number of incident deuterons is determined by SIMNRA. With that information we can calculated the Yield function, *Y(E)*.

$$Y(E) = \frac{number\ of\ Aluminium\ 26\ produced}{number\ of\ incident\ particle}$$

Figure 7 shows this function as calculated with data from each of the two monitor detectors used during deuteron irradiation.

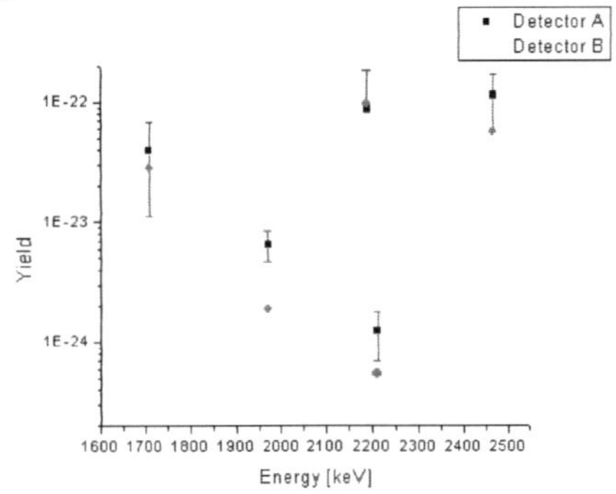

Figure 7. Shows Yield function as a function of energy.

The behavior in Yield function reflects the problems already mentioned in the derivation of both the number of ^{26}Al produced during irradiation, as well as the number of incident deuterons.

4. Summary and conclusions

This is a progress report of an ongoing effort to measure the ^{28}Si(d,α)^{26}Al excitation function at energies below the Coulomb barrier. Here we describe a number of sources of systematic errors encountered along our work. Some are related to the quantification of the number of incident beam particles on target during deuteron irradiation and others to the determination of the number of ^{26}Al produced.

We learn how to fine-tune our beam-target alignment during deuteron irradiations at the CN-Van de Graaff accelerator facility. We prove that the simple mixture Al/Si works fine for the ^{26}Al counting at our AMS facility, provided extra care is taken to reduce background from mass 28. No need for any additional chemical treatment of the irradiated substance. Having identified the main sources or systematic errors in our experimental setups, we are ready to collect new data and generate a yield function from which we could extract the desired excitation function and extend it to lower energies until we reach the practical limit of this technique, we estimate it to be near 10 nb.

It is worth to pursue this effort to produce needed data in this region, which has implications for stellar evolution models.

Acknowledgments

Special thanks to M. Veytia, R. Gleason and H. Vargas for help with design and construction at IFUNAM's mechanical shop. This work was possible thanks to financial support from CONACYT through grants numbers: 51600, 82692,123655 and DGAPAUNAM: IN-118310, IN-112609, IN-105510, IN-115213, TA-100213, IN-103312, IA-101616.

References

[1] Firestone R B, Shirley V S, and Baglin C M 1996 *Table of isotopes* Wiley Interscience
[2] Samworth E A, Warburton E K, and Engelbertink G A P 1972 *Phys. Rev. C* **5** 138
[3] Çalık A E, Gerçeklioğlu M, and Salamov D I 2009 *Zeitschrift für Naturforsch. A* **64** 865
[4] Tuniz C 2001 *Radiat. Phys. Chem.* **61** 317
[5] Suades M, Hernanz M, and De Séréville N 2010 *Highlights of Spanish Astrophysics V* Springer
[6] Lee T, Papanastassiou D A, and Wasserburg G J 1976 *Geophys. Res. Lett.* **3** 41
[7] Mahoney W A, Ling J C, Jacobson A S, and Lingenfelter R E 1982 *Astrophys. J.* **262** 742
[8] Straniero O, Imbriani G, Strieder F, Bemmerer D, Broggini C, Caciolli A, Corvisiero P, Costantini H, Cristallo S, DiLeva A, Formicola A, Elekes Z, Fülöp Z, Gervino G, Guglielmetti A, Gustavino C, Gyürky G, Junker M, Lemut A, Limata B, Marta M, Mazzocchi C, Menegazzo R, Piersanti L, Prati P, Roca V, Rolfs C, Rossi Alvarez C, Somorjai E, Terrasi F, and Trautvetter H P 2013 *Astrophys. J.* **763** 100
[9] Formicola A, Best A, Imbriani G, Junker M, Bemmerer D, Bonetti R, Broggini C, Caciolli A, Confortola F, Corvisiero P, Costantini H, Elekes Z, Fulop Z, Gervino G, Guglielmetti A, Gyürky G, Gustavino C, Lemut A, Limata B, Marta M, Mazzocchi C, Menegazzo R, Prati P, Roca V, Rolfs C, Alvarez C R, Somorjai E, Straniero O, Strieder F, Terrasi F, and Trautvetter H P 2008 *J. Phys. G: Nucl. Part. Phys.* **35** 014013
[10] Arazi A, Faestermann T, Niello J O F, Knie K, Korschinek G, Poutivtsev M, Richter E, Rugel G, and Wallner A 2006 *Phys. Rev. C* **74** 025802
[11] Strieder F, Limata B, Formicola A, and Imbriani G 2012 *Phys. Lett B* **707** 60
[12] Fitoussi C, Duprat J, and Tatischeff V 2008 *Phys. Rev. C* **78** 044613
[13] Niello J O F, Arazi A, Faestermann T, Knie K, Korschinek G, Richter E, Rugel G, and Wallner A 2003 *Brazilian J. Phys.* **33** 218
[14] Araujo-Escalona V, Andrade E, Barrón-Palos L, Canto C, Favela F, Huerta A, De Lucio O, Ortiz M E, Solís C, and Chávez E 2015 *AIP Conf. Proc* **1671** 030003
[15] Nuclear C 1993 "Alpha PIPS detectors -properties and applications," *Appl. Note*
[16] Egorov O B, Addleman R S, O'hara M J, Marks T, and Grate J W 2005 *Nucl. Instruments Methods Phys. Res. Sect. A Accel. Spectrometers, Detect. Assoc. Equip.* **537** 600
[17] Mayer M 1997 SIMNRA User's Guide, Tech. Report IPP, 9/113 *Rep. IPP*, p. 9/113
[18] Araujo-Escalona V, Acosta L, Andrade E, Cantó C, Favela F, Huerta A, De Lucio O, Solís C, and Chávez E 2016 *Nucl. Instruments Methods Phys. Res. Sect. B Beam Interact. with Mater. Atoms*, under review

Journal of Physics: Conference Series **730** (2016) 012004 doi:10.1088/1742-6596/730/1/012004

Recent Nuclear Astrophysics Measurements using the *TwinSol* Separator

D W Bardayan[1], T Ahn[1], J Allen[1], F D Becchetti[2], J C Blackmon[3], M Brodeur[1], B Frentz[1], Y K Gupta[1,4], M R Hall[1], O Hall[1,5], S Henderson[1], J Hu[1,6], J M Kelly[1], J J Kolata[1], A Long[1], J Long[1], K Macon[3], C Nicoloff[1,7], P D O'Malley[1], K Ostdiek[1], S D Pain[8], J Riggins[2], B E Schultz[1], M Smith[1], S Strauss[1], R O Torres-Isea[2]

[1]Department of Physics, University of Notre Dame, Notre Dame, IN 46556, USA
[2]Physics Department, University of Michigan, Ann Arbor, MI 48109, USA
[3]Department of Physics and Astronomy, Louisiana State University, Baton Rouge, LA 70803, USA
[4]Nuclear Physics Division, Bhabha Atomic Research Centre, Mumbai, 400085, India
[5]Department of Physics, University of Surrey, Guildford, GU2 7XH, United Kingdom
[6]Institute of Modern Physics, Chinese Academy of Sciences, Lanzhou 730000, China
[7]Department of Physics, Wellesley College, Wellesley, MA 02481, USA
[8]Physics Division, Oak Ridge National Laboratory, Oak Ridge, TN 37831, USA

E-mail:danbardayan@nd.edu

Abstract. Many astrophysical events, such as novae and X-ray bursts, are powered by reactions with radioactive nuclei. Studying the properties of these nuclei in the laboratory can therefore further our understanding of these astrophysical explosions. The *TwinSol* separator at the University of Notre Dame has recently been used to produce intense (~10^6 pps) beams of ^{17}F. In this article, some of the first measurements with these beams are discussed.

1. Introduction

Nova explosions and X-ray bursts are brilliant astrophysical events resulting from the rapid accretion of hydrogen-rich material onto a degenerate star from a binary companion [1]. Convection mixes underlying material with this layer of hydrogen, and thermonuclear reactions are initiated as the temperatures and densities rise. These reactions produce radioactive proton-rich nuclei, which can be involved in further nuclear processing owing to the short reaction times in such environments. The explosion dynamics and nucleosynthesis in such explosions are therefore strongly dependent upon the reaction rates and the properties of the proton-rich, exotic nuclei produced.

2. Production of Radioactive Beams

At the Notre Dame Nuclear Science Laboratory (NSL), studies of proton-rich nuclei are enabled via the use of the *TwinSol* magnetic separator [2]. Intense (~ 1 μA) stable beams are accelerated up to ~5 MeV/u by the NSL 11-MV FN tandem accelerator for bombardment of light-ion gas targets to produce exotic nuclei. The use of inverse-kinematical production reactions results in higher-energy, forward-moving radioactive beams and higher-efficiency capture of the exotic products by the *TwinSol* solenoids. The production target consists of 4 independent gas cells that typically contain ~1 atmosphere of hydrogen or helium gas held by 5-μm-thick Ti windows. A picture of the production target assembly is shown in Fig. 1. The use of multiple gas cells on a single motorized feedthrough allows for fast and efficient changeovers in the event of window deterioration and rupture. The primary beam is subsequently stopped and monitored by a Faraday cup directly behind the target while the exotic reaction products produced at angles between 2-5 degrees are transmitted through the *TwinSol* device. Within the past year, a primary focus of the TwinSol nuclear astrophysics program has been the production and use of intense ^{17}F beams, the discussion of which comprises the remainder of this article.

Fig. 1: The new NSL multi-cell gas production target is shown in the initial vacuum testing stage with blank flanges (left) along with the windows installed (right).

^{17}F beams between 40-52 MeV are produced by bombarding deuterium gas with ^{16}O beams between 64-74 MeV. The energy range of exotic beam products is determined by the reaction kinematics of the ^{2}H(^{16}O,^{17}F)n reaction and by the energy loss in the gas-cell and windows. The secondary ^{17}F beam is subsequently purified by separation and focusing with the *TwinSol* solenoids onto a secondary target ~2 m downstream from the second solenoid. The beam composition is assayed at the secondary target position via use of a silicon surface barrier detector telescope and a position-sensitive ionization counter (see Fig. 2). The secondary beam is primarily composed of ^{17}F and ^{16}O with ratios varying between ^{17}F/^{16}O ~ 0.5-1.5 depending upon the particular tune. In general the ^{16}O contaminants are not well focused at the secondary target positions, and thus the beam purity can be improved by appropriate beam collimation. Typical collimation size is between 0.5-2.5 cm depending upon the type of measurement being performed. Trace amounts of ^{17}O, ^{15}O, ^{14}N, and ^{12}C are also observed in the beam. Secondary beam intensities of up to 2×10^{6} ^{17}F/s are observed for primary beam intensities of 1 μA of q=+7 ^{16}O.

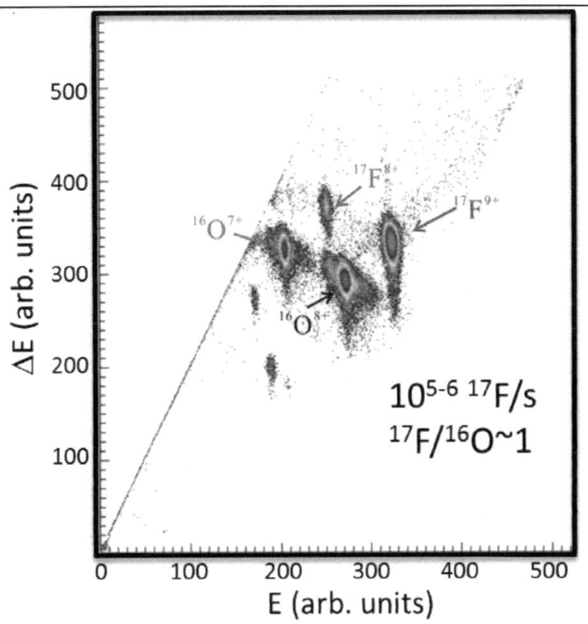

Fig. 2: Composition of the secondary beam as sampled by a ΔE-E silicon detector telescope. The beam is primarily composed of ^{17}F and ^{16}O.

3. ^{17}F half-life measurement

In astrophysical explosions, one of the primary determinants of the nucleosynthesis is the lifetimes of the nuclei involved. Whether a synthesized radioactive nucleus is subject to further nuclear reactions or simply decays back towards stability is largely determined by its half-life. ^{17}F is produced in large amounts in novae, but despite years of study, there was not good agreement for the half-life of ^{17}F (as shown in Table 1), and thus a new measurement was warranted.

Table 1: Previous ^{17}F half-life measurements are shown.

^{17}F Half-life (s)	Reference
66.0±0.2	[3]
65.2±0.2	[4]
64.50±0.25	[5]
64.8±0.12	[6]
64.31±0.09	[7]

The secondary ^{17}F beam from *TwinSol* was implanted into thick tantalum foils on the end of a rotating arm. After a fixed implantation time, the foil was rotated 180 degrees to face a 25-mm-thick plastic scintillator coupled to a photomultiplier tube. While one foil was being counted by the plastic scintillator, a second foil would be subject to further implantation. After the same fixed amount of time, the rotation and counting sequence would begin again. The number of counts observed in the plastic scintillator were referenced to an accurate 100 Hz clock that only deviated by approximately 3s in 10 h during a calibration test. The data collected by this procedure is plotted in Fig. 3 along with a decay-curve fit taking into account the small amount of ^{15}O ($^{15}O/^{17}F\sim10^{-3}$) in the beam and letting the background detector count rate (~2 Hz from room γ background) vary as a parameter of the fit. The half-life resulting from this fit was 64.402 ± 0.042 s [8], which agrees well with another recent measurement, 64.347 ± 0.035 s [9] but not several earlier measurements (Table 1).

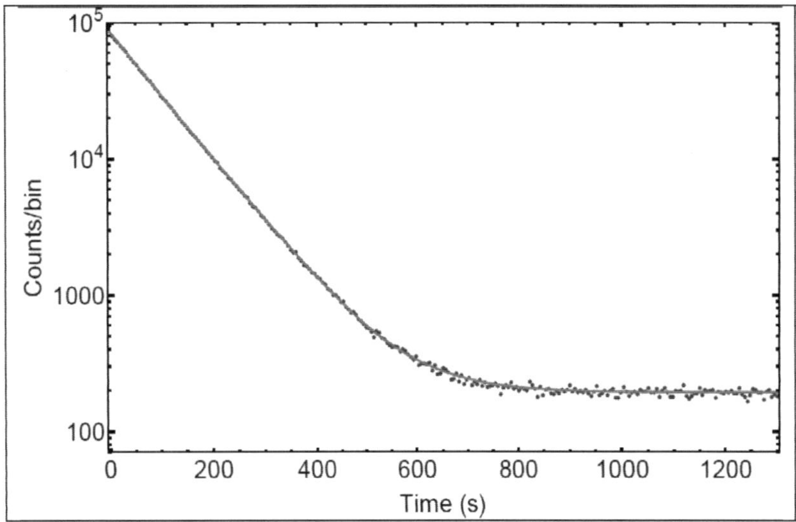

Fig 3: The number of decay events observed as a function of time in 5-second bins from several cycles of ^{17}F implantation into Ta foils, rotation in front of a plastic scintillator, and counting β particles from the decay.

4. Study of the $^{17}F(p,\alpha)^{14}O$ reaction

A second thrust of the work with ^{17}F beams is to better understand the $^{14}O(\alpha,p)^{17}F$ reaction rate in X-ray bursts. Owing to the difficulty in studying the reaction rate directly, the time-inverse reaction, $^{17}F(p,\alpha)^{14}O$, will be studied instead. States in ^{18}Ne from E_x=6-7.5 MeV provide important resonances in the $^{14}O(\alpha,p)^{17}F$ reaction rate at X-ray burst temperatures, and determining their properties is a top priority [10]. A number of studies have been performed of the $^{17}F(p,\alpha)^{14}O$ reaction [10-12] and the existing cross section data is plotted in Fig. 4. While these data seem adequate above ^{17}F beam energies of 55 MeV, below this energy the measurements are sparse and suffer from large uncertainties, making it difficult to assess any possible resonance structure. Particularly important is determining the properties of potential resonances near $E(^{17}F)$=40 MeV that could correspond to a known ^{18}Ne state at E_x=6.15 MeV [10].

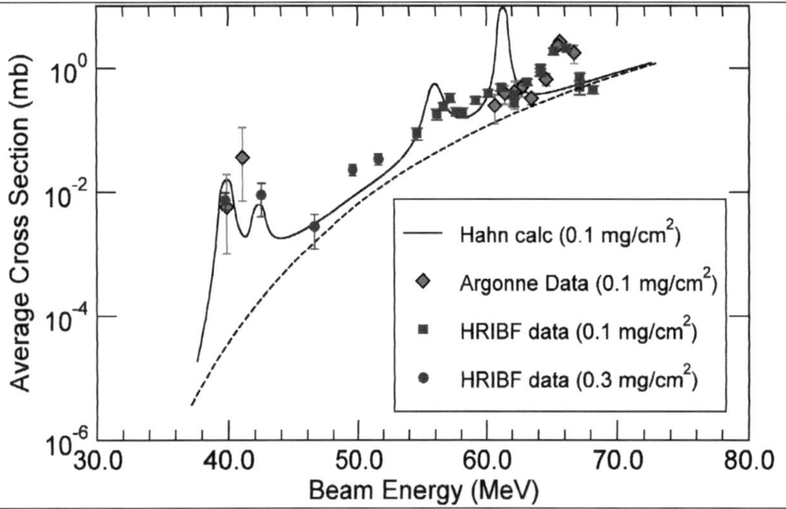

Fig. 4: Previous $^{17}F(p,\alpha)^{14}O$ data sets are shown with the legend indicating the CH_2 target thickness that was used in each case. The calculation uses resonance parameters from Hahn *et al.* [10] to estimate the expected cross section averaged over the energy loss for a 0.1 mg/cm^2 target.

To measure the $^{17}F(p,\alpha)^{14}O$ cross section below 55 MeV, polyethylene $(CH_2)_n$ foils will be bombarded with intense ^{17}F beams produced by *TwinSol*. The reaction products, ^{14}O ions and α particles, will be detected in coincidence in annular arrays of silicon strip detectors. Events from the $^{17}F(p,\alpha)^{14}O$ reaction can be distinguished from other reactions products by their distinctive reaction kinematics and by requiring a tight time-correlation between the detected ^{14}O ions and α particles. It is envisioned that further clarification of the data can be accomplished via pulsing the beam in short (1-2 ns) bunches and measuring the time-of-flight of ions through the *TwinSol* separator. This should help reduce background events from the contamination of other ions in the secondary beam.

As a test of the procedure and setup, a measurement of the $^{17}O(p,\alpha)^{14}N$ reaction was performed at *TwinSol*. In this case, the production target and Faraday cup were removed so that the ^{17}O beam could be directly tuned through the separator. The beam then bombarded a 400 μg/cm^2 polyethylene foil and reaction α particles were detected at angles between 8-24 degrees in coincidence with recoil ^{14}N ions. By requiring a proper time coincidence and Q-value for the reaction, the events in Fig. 5 were selected. The events from the $^1H(^{17}O,\alpha)^{14}N$ and $^1H(^{17}O,p)^{17}O$ reactions were clearly visible and separated. After this successful test run, future studies will focus on measuring the $^{17}F(p,\alpha)^{14}O$ cross section directly.

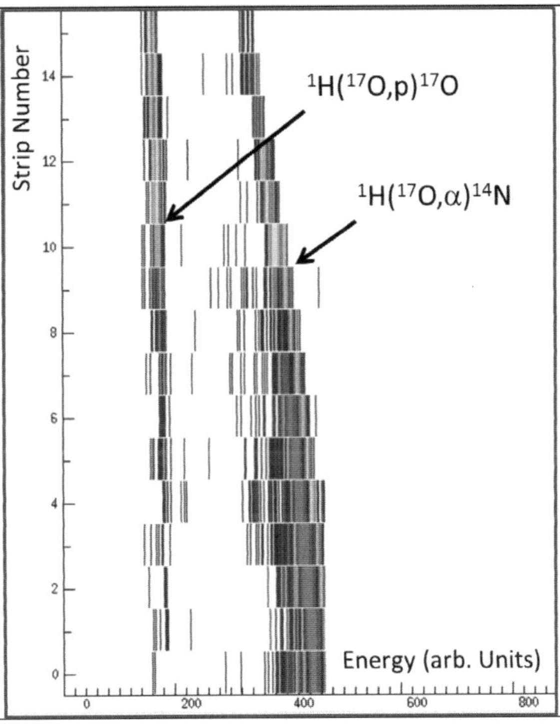

Fig. 5: The strip number of detected events is plotted against the observed energy for events satisfying gates on the expected Q-value and proper time coincidence. The angles covered start at 8 degrees for strip 0 and go to 24 degrees for strip 15. Using these cuts, events from the $^1H(^{17}O,\alpha)^{14}N$ reaction are clearly resolved.

5. Conclusions

In conclusion, the availability of exotic beams is opening new and unprecedented opportunities to study astrophysical explosions such as novae and X-ray bursts. *TwinSol* at the University of Notre Dame was the first such device in the U.S. dedicated to the production of radioactive beams. Recent upgrades to *TwinSol* have improved its ability to study nuclei and reactions of astrophysical interest. The uncertainty in the ^{17}F half-life has been greatly reduced, and preparations are on-going for a new study of the $^{14}O(\alpha,p)^{17}F$ reaction.

This research was supported by the Department of Energy Office of Nuclear Physics, the National Nuclear Security Agency, and the National Science Foundation.

[1] Starrfield S, Truran J W, Sparks W M and Kutter G S 1972 *Astrophys. J.* **176** 169.
[2] Lee M Y *et al.* 1997 *AIP Conf. Proc.* **392** 397.
[3] Janecke V J 1960 *Z. Naturforschg.* **15 a** 593.
[4] Wohllenben V K and Schuster E 1969 *Radiochimica Acta* **12** 75.
[5] Alburger D E and Wilkinson D H 1972 *Phys. Rev. C* **6** 2019.
[6] Alburger D E 1977 *Phys. Rev. C* **16** 889(R).
[7] Azuelos G, Kitching J E, and Ramavataram K 1977 *Phys. Rev. C* **15** 1847.
[8] Brodeur M *et al.* 2016 *Phys. Rev. C* **93** 025503.
[9] Grinyer J *et al.* 2015 *Phys. Rev. C* **92** 045503.
[10] Hahn K I *et al.* 1996 *Phys. Rev. C* **54** 1999.
[11] Harss B *et al.* 1999 *Phys. Rev. Lett.* **82** 3964.
[12] Blackmon J C *et al.* 2001 *Nucl. Phys. A* **688** 142c.

Journal of Physics: Conference Series **730** (2016) 012005 doi:10.1088/1742-6596/730/1/012005

Strangeness suppression in the unquenched quark model

Roelof Bijker
Instituto de Ciencias Nucleares, Universidad Nacional Autónoma de México, A.P. 70-543, 04510 México, D.F., México
E-mail: bijker@nucleares.unam.mx

Hugo García-Tecocoatzi
Instituto de Ciencias Nucleares, Universidad Nacional Autónoma de México, A.P. 70-543, 04510 México, D.F., México
I.N.F.N., Sezione di Genova, via Dodecaneso 33, I-16146 Italy
E-mail: hugo.garcia@nucleares.unam.mx

Elena Santopinto
I.N.F.N., Sezione di Genova, via Dodecaneso 33, I-16146 Italy
E-mail: elena.santopinto@ge.infn.it

Abstract. In this contribution, we discuss the strangeness suppression in the proton in the framework of the unquenched quark model. The theoretical results are in good agreement with the values extracted from CERN and JLab experiments.

1. Introduction

The importance of multiquark configurations in the proton such as $qqq - q\bar{q}$ has been inferred from the observed violation of the Gottfried sum rule, that is that there are more anti-d than anti-u quarks in the proton, and the proton spin crisis in which only about one third of the proton spin is carried by the quark spins. Recently, interesting new results were reported by the CLAS Collaboration [1], in which the strangeness suppression in the proton was determined from the production rates of baryon-meson states in exclusive reactions, *i.e.* without the production of an intermediate baryon resonance.

The purpose of this contribution is study the observed production rates of meson-baryon final states and the strangeness suppression in the proton in the framework of an unquenched quark model.

2. Strangeness suppression

In the unquenched quark model (UQM) the effects of quark-antiquark pairs are introduced via a $^3P_0$ mechanism [2, 3, 4, 5, 6, 7]. The production ratios of baryon-meson states can be expressed

Table 1. Ratios of electroproduction cross sections.

Ratio	UQM [8]	Exp. [1]
$p \to \Lambda K^+/p \to n\pi^+$	0.227	$0.19 \pm 0.01 \pm 0.03$
$p \to \Lambda K^+/p \to p\pi^0$	0.454	$0.50 \pm 0.02 \pm 0.12$
$p \to p\pi^0/p \to n\pi^+$	0.500	$0.43 \pm 0.01 \pm 0.09$

Table 2. Strangeness suppression in the proton.

Ratio	UQM [8]	Exp.	Ref.
$s\bar{s}/d\bar{d}$	0.265	0.22 ± 0.07	[1]
$u\bar{u}/d\bar{d}$	0.568	0.74 ± 0.18	[1]
$2s\bar{s}/(u\bar{u} + d\bar{d})$	0.338	0.25 ± 0.08	[1]
		0.29 ± 0.02	[9]

as the product of a spin-flavor-isospin factor and a radial integral [8]

$$\frac{p \to \Lambda K^+}{p \to n\pi^+} = \frac{27}{50}\frac{I_{N\to\Lambda K}}{I_{N\to N\pi}} , \tag{1}$$

with

$$I_{A\to BC} = \int_0^\infty \frac{k^4 e^{-2F^2k^2}}{\Delta E^2_{A\to BC}(k)} dk . \tag{2}$$

The energy denominator represents the energy difference between initial and final hadrons calculated in the rest frame of the initial baryon A. The value of F^2 depends on the size of the harmonic oscillator wave functions for baryons and mesons and the Gaussian smearing of the pair-creation vertex, and is taken from [7] to be $F^2 = 2.275$ GeV^{-2}. Table 1 shows that the observed rates are reproduced very well by our calculation. The calculated ratio $p \to p\pi^0/p \to n\pi^+ = 1/2$ is a consequence of the isospin symmetry of the UQM.

In Ref. [1] a simple factorization model was proposed to extract the ratio of $q\bar{q}$ probabilities. In the UQM the corresponding ratios for exclusive two-body production can be determined in a straightforward way. The present calculation takes into account all channels involving pseudoscalar mesons (π, K, η and η') in combination with octet and decuplet baryons. The results are presented in Table 2. The value for the strangeness suppression $\lambda_s = 2s\bar{s}/(u\bar{u}+d\bar{d})$ is in good agreement with both the values determined in exclusive reactions [1] and in high-energy production [9].

3. Summary and conclusions

In conclusion, the observed ratios for the production of baryon-meson channels in exclusive reactions can be understood in a simple and transparent way in the framework of the UQM. It is important to emphasize that the UQM results do not depend on the strength of the $^3P_0$ quark-antiquark pair creation vertex, The value of the remaining coefficient (F^2) was taken from previous work, no attempt was made to optimize their values. Finally, the UQM value for the strangeness suppression in the proton is in good agreement with the value determined in exclusive reactions [1] as well as the result from high-energy production [9].

Acknowledgments
This work was supported in part by research grant IN107314 from PAPIIT-DGAPA, Mexico and in part by INFN, Sezione di Genova, Italy.

References
[1] Mestayer M D *et al.* 2014 *Phys. Rev. Lett.* **113** 152004
[2] Törnqvist N A and Zenczykowski P 1986 *Z. Phys. C* **30** 83
[3] Zenczykowski P 1986 *Ann. Phys. (N.Y.)* **169** 453
[4] Geiger P and Isgur N 1997 *Phys. Rev. D* **55** 299
[5] Bijker R and Santopinto E 2009 *Phys. Rev. C* **80** 065210
[6] Santopinto E and Bijker R 2010 *Phys. Rev. C* **82** 062202(R)
[7] Bijker R, Ferretti J and Santopinto E 2012 *Phys. Rev. C* **85** 035204
[8] Santopinto E, Bijker R and García-Tecocoatzi H 2016 *Phys. Lett. B* **759** 214 [arXiv:1601.06987].
[9] Bocquet G *et al.* 1996 *Phys. Lett. B* **366** 447

Journal of Physics: Conference Series **730** (2016) 012006 doi:10.1088/1742-6596/730/1/012006

The nuclear matrix elements of 0νββ decay and the NUMEN project at INFN-LNS

F. Cappuzzello[1,2,•], C. Agodi[1,•], E. Aciksoz[14], L. Acosta[11], X. Aslanouglou[10], N. Auerbach[16], R. Bijker[11], D. Bonanno[3], D. Bongiovanni[3], T. Borello[8], S .Boudhaim[15], M.L. Bouhssa[15], I. Boztosun[14], L. Calabretta[1], A. Calanna[1], D. Carbone[1], M. Cavallaro[1], D. Calvo[4], E.R. Chávez Lomelí[11], M. Colonna[1], G. D'Agostino[1,2], N. Deshmukh[1], P.N. de Faria[9], A. Ferrero[4,6], A. Foti[2,3], P. Finocchiaro[1], P.R.S. Gomes[9], V. Greco[1], A. Hacisalihoglu[14], Z. Housni[15], A. Khouaja[15], J. Inchaou[15], G. Lanzalone[1,7], F. La Via[1,12], J.A. Lay[1], H. Lenske[13], R. Linares[9], J. Lubian[9], F. Iazzi[4,6], R. Introzzi[4,6], A. Lavagno[4], D. Lo Presti[2,3], N. Medina[8], D. R. Mendes[9], A. Muoio[1], J.R.B. Oliveira[8], A. Pakou[10], L. Pandola[1], D. Rifuggiato[1], M.R.D. Rodrigues[8], G. Santagati[1], E. Santopinto[5], L. Scaltrito[4], O. Sgouros[10], S.O. Solakcı[14], V. Soukeras[10], S. Tudisco[1], R.I.M. Vsevolodovna[5], V. Zagatto[8]

1. *Istituto Nazionale di Fisica Nucleare, Laboratori Nazionali del Sud, Catania, Italy*
2. *Dipartimento di Fisica e Astronomia, Università di Catania, Italy*
3. *Istituto Nazionale di Fisica Nucleare, Sezione di Catania,*
4. *Istituto Nazionale di Fisica Nucleare, Sezione di Torino*
5. *Istituto Nazionale di Fisica Nucleare, Sezione di Genova*
6. *Politecnico di Torino, Italy*
7. *Università degli Studi di Enna "Kore", Enna, Italy*
8. *Universidade de Sao Paulo, Brazil*
9. *Universidade Federal Fluminense, Niteroi, Brazil*
10. *University of Ioannina, Ioannina, Greece*
11. *Universidad Nacional Autónoma de México*
12. *CNR-IMM, Sezione di Catania, Italy*
13. *University of Giessen, Germany*
14. *Akdeniz University, Antalya, Turkey*
15. *Université Hassan II – Casablanca, Morocco*
16. *School of Physics and Astronomy Tel Aviv University, Israel*

Abstract

An innovative technique to access the nuclear matrix elements entering the expression of the life time of the double beta decay by relevant cross section measurements of double charge exchange reactions is proposed. A key aspect of the project is the use of the MAGNEX large acceptance magnetic spectrometer, for the detection of the ejectiles, and of the LNS K800 Superconducting Cyclotron (CS), for the acceleration of the required high resolution and low emittance heavy-ion beams, already in operation at INFN Laboratory Nazionali del Sud in Catania (Italy). However, a major upgrade is foreseen for the INFN-LNS research infrastructure to cope with beam currents as high as several pμA required by the project.

1 Introduction

Neutrinoless double beta decay, 0νββ, is at the present time strongly pursued both experimentally and theoretically [1]. Its observation will determine whether the neutrino is a Dirac or Majorana particle and will provide a measurement of the average neutrino mass, which is nowadays

• Spokespersons: F. Cappuzzello (cappuzzello@lns.infn.it) and C. Agodi (agodi@lns.infn.it)

one of the fundamental problems in physics. An innovative technique to access the Nuclear Matrix Elements (NME) entering the expression of the life time of the 0νββ by relevant cross sections of double charge exchange (DCE) reactions is proposed. The basic point is the coincidence of the initial and final state wave-functions in the two classes of processes and the similarity of the transition operators, which in both cases present a superposition of Fermi, Gamow-Teller and rank-two tensor components with a relevant momentum available.

First pioneering experimental results, obtained at the INFN-LNS laboratory in Catania, for the ^{40}Ca(^{18}O,^{18}Ne)^{40}Ar reaction at 270 MeV, in a wide range of transferred momenta, give encouraging indication on the capability to access quantitative information towards the determination of the NME for 0νββ decay [2]. On the basis of the above mentioned ground-breaking achievement, we propose an ambitious project, NUMEN, with the aim to go deep insight in the Heavy Ion DCE studies on nuclei of interest for 0νββ decay [3].

There are a number of important similarities among DCE and 0νββ decay processes, despite they are mediated by different interactions. The description of NMEs extracted from DCE and 0νββ presents the same degree of complexity, with the advantage for DCE to be "accessible" in laboratory. However, a simple relation between DCE cross sections and ββ-decay half-lives is not trivial and needs to be explored.

1.1 The Project

The availability of the MAGNEX spectrometer [4] for high resolution measurements of very suppressed reaction channels was essential for the first pilot experiment. Moreover, the measurement of DCE high resolution energy spectra and accurate cross sections at very forward angles are key points to identify the transitions of interest [5]. The concurrent measurement of the other relevant reaction channels allows to isolate the direct DCE mechanism from the competing transfer processes. These are at least of 4th-order and can be effectively minimized by the choice of the proper projectile-target system and incident energy [6].

Fig.1: A view of MAGNEX spectrometer at Laboratori Nazionali del Sud in Catania.

Fig.2: A view of K800 Superconducting Cyclotron at Laboratori Nazionali del Sud in Catania.

The present limit of low beam current we have experienced both for the CS accelerator and for the MAGNEX focal plane detector must be sensibly overcome. For a systematic study of the many "hot" cases of ββ decays an upgraded set-up, able to work with two orders of magnitude more current than the present, is thus necessary. This goal can be achieved by a substantial change in the technologies used in the beam extraction and in the detection of the ejectiles. For the accelerator the use of a stripper induced extraction is an adequate choice. However, with the present set-up it is difficult to suitably extend this research to the "hot" cases, where ββ decay studies are and will be concentrated. For the spectrometer the main foreseen upgrades are:

1. The substitution of the present Focal Plane Detector (FPD) [7-8] technology with tracker based on a micro-pattern electron amplifier;

2. The substitution of the wall of silicon pad stopping detectors with a wall of telescopes based on Silicon Carbide detectors;

3. The enhancement of the maximum magnetic rigidity;

4. The introduction of an array of detectors for measuring the coincident γ-rays.

In this framework we propose four phases in the NUMEN project, looking forward to do, in the same time, both the experimental and the up-grade activity, as indicated in the following Phases of the project.

Phase1: the experiment feasibility

The pilot experiment: $^{40}Ca(^{18}O,^{18}Ne)^{40}Ar$ reaction at 270 MeV, with the first experimental data on heavy-ion DCE reactions in a wide range of transferred momenta, was already done. The results demonstrate the technique feasibility.

Phase2: toward "hot" cases optimizing experimental conditions and getting first results

The necessary work for the upgrading of both the accelerator and MAGNEX will be carried out still preserving the access to the present facility. Due to the relevant technological challenges connected, in which test, with and without beam will be crucial, the Phase2 is foreseen to have a duration of a 3-4 years. In the meanwhile, experiments with integrated charge of tens of mC (about one order of magnitude more than that collected in the pilot experiment) will be performed. These will require several weeks (4-8 depending on the case) data taking for each reaction, since thin targets (a few 10^{18} atoms/cm^2) are mandatory in order to achieve enough energy and angular resolution in the energy spectra and angular distributions. The attention will be focused on a few favorable cases, like for example ^{116}Sn (^{18}O,^{18}Ne)^{116}Cd reaction at 15 and 30 MeV/u and the ^{116}Cd (^{20}Ne,^{20}O)^{116}Sn reaction at 15 and 25 MeV/u, with the goal to achieve conclusive results for them.

Phase3: the facility upgrade

Once all the building block for the upgrade of the accelerator and spectrometer facility will be ready at the LNS a Phase3, connected to the disassembling of the old set-up and re-assembling of the new will start. An estimate of about 18-24 months is considered.

Phase4: the experimental campaign

The Phase4 will consist of a series of experimental campaigns at high beam intensities (some pμA) and long experimental runs in order to reach in each experiment integrated charge of hundreds of mC up to C, for the experiments in coincidences, spanning all the variety of candidate isotopes for *0νββ* decay, like:

^{48}Ca, ^{82}Se, ^{96}Zr, ^{100}Mo, ^{110}Pd, ^{124}Sn, ^{128}Te, ^{130}Te, ^{136}Xe, ^{148}Nd, ^{150}Nd, ^{154}Sm, ^{160}Gd, ^{198}Pt.

1.1.1 The NUMEN goals

Studying if the DCE reaction mechanism can be reliably factorized in a reaction part and a nuclear structure part, this latter factorized in a projectile and target matrix element is a main purpose of NUMEN. The development of a consistent microscopic description of the heavy-ion double charge exchange reaction and the nuclear structure part is essential to this purpose. The use of the quantum approach for the DWBA or CRC cross section with form factors including QRPA transition densities (as well as Shell Model or IBM densities) is a suitable framework in which this theory can be developed. Experimentally the achievement of the first goal requires that a systematic set of appropriate data is built, facing the relative experimental challenges connected with the low cross sections and high resolutions requests.

Moreover, the project has two other goals, ground-breaking and achievable in a shorter period. The measured DCE absolute cross sections provide themselves a powerful tool for tuning the nuclear structure theory. The matrix elements for double charge exchange and neutrino-less double beta decay probe the same initial and final wave functions by operators with similar structure. Consequently, the measured DCE absolute cross sections allows to test the goodness of the assumptions done for the unavoidable truncation of the many-body wave functions. The reaction part need to be precisely controlled to this purpose, a result that is at reach within a fully quantum scattering framework. Once the nuclear wave functions have been tested by DCE cross sections, the same can be used for 0*νββ* decay nuclear matrix elements. Promoting the development of these kind of constrained theories for the NME of the neutrino-less double beta decay is thus an important goal that NUMEN can achieve even with a reduced experimental dataset and without assuming cross section factorization.

Finally, the third goal is to provide relative NME information on the "hot cases" of the 0*νββ* decay. In case of validity of cross section factorization, the ratio of measured cross sections can give a model independent way to compare the sensitivity of different half-life experiments. This result can be achieved even in presence of sizeable systematic errors in the measured cross sections and in the extraction of unit cross sections, as they are largely reduced in the ratio. Performing these comparative analyses could strongly impact in the future developments of the field, especially in a scenario were fundamental choices for the best isotope candidates for $0\nu\beta\beta$ decay need to be done.

References

[1] Report to the Nuclear Science Advisory Committee, Neutrinoless Double Beta Decay, 2015

[2] Cappuzzello F, Cavallaro M *et al.* 2015 *Eur. Phys. J. A* **51** 145

[3] Agodi C *et al.* 2015 *Nuclear and Particle Physics Proceedings* **265-266** 28

[4] Cappuzzello F *et al.* 2011 *MAGNEX: an innovative large acceptance spectrometer for nuclear reaction studies in: Magnets: Types, Uses and Safety* (New York: Nova Publisher Inc.) 1

[5] Matsubara H *et al.* 2013 *Few-Body Systems* **54** 1433

[6] Brink D M 1972 *Phys. Lett. B* **40** 37

[7] Cavallaro M *et al.* 2012 *Eur. Phys. J. A* **48** 59

[8] Carbone D *et al.* 2012 *Eur. Phys. J. A* **48** 60

Journal of Physics: Conference Series **730** (2016) 012007 doi:10.1088/1742-6596/730/1/012007

The Giant Pairing Vibration in Carbon isotopes

M. Cavallaro[a*], F. Cappuzzello[a,b], D. Carbone[a], C. Agodi[a], F. Azaiez[c], M. Assié[c], N. de Séréville[c], J. A. Scarpaci[d]

[a] Istituto Nazionale di Fisica Nucleare, Laboratori Nazionali del Sud, I-95125 Catania, Italy.

[b] Dipartimento di Fisica e Astronomia, Università di Catania, I-95125 Catania, Italy.

[c] Institut de Physique Nucléaire, Université Paris-Sud-11-CNRS/IN2P3, 91406 Orsay, France

[d] Centre de Sciences Nucléaires et de Sciences de la Matières - CSNSM, Université Paris-Sud-11-8 CNRS/IN2P3, 91405 Orsay, France

*manuela.cavallaro@lns.infn.it

Abstract. The $^{13}C(^{18}O,^{16}O)^{15}C$ and $^{12}C(^{18}O,^{16}O)^{14}C$ reactions at 84 MeV incident energy were explored up to high excitation energy of the residual nucleus thanks to the use of the MAGNEX spectrometer to detect the ejectiles. In the region above the two-neutron separation energy, a resonance has been observed in both nuclei, attributed to the Giant Pairing Vibration (GPV). The neutron decay of the ^{15}C resonances, including the GPV, populated via the two-neutron transfer reaction has been studied using an innovative technique, which couples MAGNEX with the EDEN neutron detector array. The data show that the ^{15}C GPV mainly decays via two-neutron emission.

1. Introduction

In the last few years, a study of the structure of different nuclei was pursued at the Catania INFN-LNS laboratory by the ($^{18}O,^{16}O$) two-neutron transfer reaction at 84 MeV on different targets using the MAGNEX large acceptance magnetic spectrometer to detect the ejectiles. Thanks to its high resolution and large acceptance, high quality inclusive spectra were obtained, even in a largely unexplored region above the two-neutron emission threshold in the residual nucleus. New phenomena appeared, such as the dominance of the direct one-step transfer of the two neutrons [1] and the presence of broad resonances at high excitation energy in the ^{14}C and ^{15}C spectra [2]. These structures were recently identified as the first experimental signature of the Giant Pairing Vibration (GPV) [3], [4], [5] predicted long time ago [6].

The GPV is a giant collective mode in the atomic nuclei. Its microscopic origin is the coherence among elementary particle-particle excitations generated by the addition or removal to/from an atomic nucleus of two nucleons in a relative S-wave (motion component characterized by an orbital angular momentum L = 0). This quantum mechanism is analogous to that feeding the Giant Resonances (GR) observed in the inelastic nuclear excitation. The link between GPV and GR is the symmetry between single-particle states above (particle) and below (hole) the Fermi level. In both cases, the only required condition is a system whose minimum energy structure is governed by a mean field. GR are collective states, manifestation of a coherence mechanism of the particle-hole (p-h) excitations connecting major shells of the harmonic oscillator like nuclear mean field. In a similar way, the GPV represents the particle-particle (p-p) or the hole-hole (h-h) counterpart of the GR. The main component of the p-p or h-h residual interaction is the pairing field. The introduction of such a field explains why at the minimum energy the nuclei with an even number of neutrons and protons are more tightly bound than the neighbouring even-odd or odd-odd ones.

Despite the existence of the GPV mode has been theoretically suggested and several evidences of p-h giant modes have been observed, like the GR, no resonances with the properties defining the GPV has been identified in any nucleus so far. This is probably due to the reaction mechanism required for populating the GPV. The addition or removal of two nucleons can be obtained in the laboratory by two-nucleon transfer reactions. These processes are typically characterized by a sizeable exchange of linear momentum at the surface of the colliding nuclei. Consequently, the angular momentum transferred in the reaction tends to be large and the cross section for the L = 0 modes is expected to be suppressed. This feature is more evident for reactions induced by heavy-ion projectiles and characterized by large negative Q-values. In Ref. [3] we reported on a new experimental approach to investigate the GPV mode, based on heavy-ion induced two-neutron transfer reactions on light nuclei.

2. Experiments and results

In the experiment conducted at INFN-LNS laboratory, a beam of $^{18}O^{6+}$ ions at 84 MeV incident energy impinged on a 49 ± 3 μg/cm^2 self-supporting ^{12}C and a 50 ± 3 μg/cm^2 self-supporting 99% enriched ^{13}C targets. The runs with ^{12}C were used also for estimating the background coming from ^{12}C impurities in the ^{13}C target. The ^{16}O ejectiles produced in the collisions were momentum analysed by the MAGNEX spectrometer working in full acceptance mode (solid angle 50 msr and momentum acceptance $\Delta p/p \sim 24\%$).

The neutron decay of the ^{15}C resonances up to 16 MeV excitation energy, populated via the $^{13}C(^{18}O,^{16}O\ n)$ reaction was studied using a new method to determine the neutron kinetic energy by time-of-flight (TOF) in exclusive experiments [7]. It involves the use of MAGNEX coupled to the EDEN neutron detector array [8], [9]. A schematic view of the used set-up is shown in Figure 1.

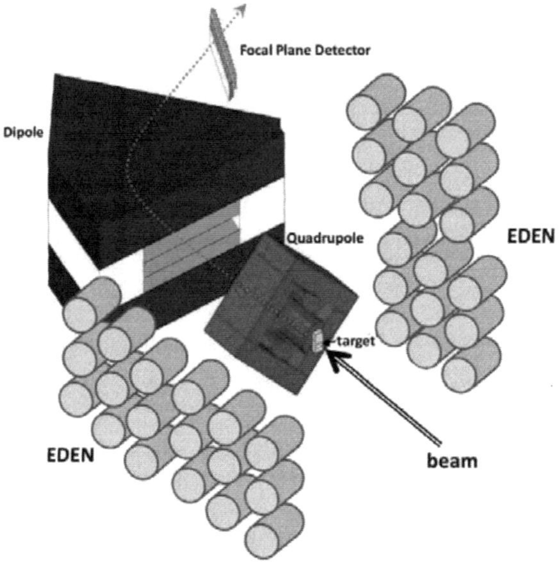

Figure 1. Schematic layout of the MAGNEX-EDEN facility.

The reconstructed excitation energy spectra $E_x = Q_0 - Q$ (where Q_0 is the ground to ground state Q-value) were obtained by the missing mass determination by MAGNEX applying an algebraic ray-reconstruction technique [10], [11]. An example for the ^{15}C case is shown in Figure 2. The background contribution, which mainly comes from impurities of ^{12}C in the ^{13}C target, was evaluated and subtracted in the spectrum. Several narrow peaks corresponding to well-known low-lying bound and resonant states are observed.

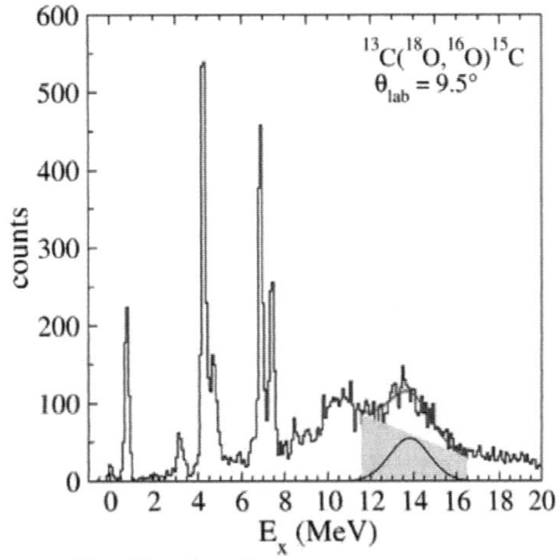

Figure 2. Energy spectra for the $^{13}C(^{18}O,^{16}O)^{15}C$ reaction. The linear background model, the fit of the GPV bump and their sum are shown as the grey area, the black Gaussian and the red line, respectively.

A sudden increase of the yield is found just above the two-neutron separation energy (S_{2n}= 9.394 MeV). In particular, a large bump, superimposed to the continuum background, is observed in the spectrum. A best fit procedure with Gaussian shapes and a locally adjusted linear background model gives an energy E_x = 13.7 ± 0.1 MeV (FWHM 1.9 ± 0.3 MeV). A similar spectrum is obtained in the case of ^{14}C, showing the presence of an unknown bump at 16.9 ± 0.1 MeV. The centroids and widths are confirmed by a supplementary experiment performed at 270 MeV bombarding energy studying the same reactions by the same experimental set-up.

The absolute cross section angular distributions of the two bumps were extracted, according to the procedure described in Refs. [12], [4]. The background underneath each peak was modelled by a linear function and the possible overlap between close peaks was treated by a least squared analysis assuming Gaussian shapes. The angular distribution for the bump at 13.7 MeV in ^{15}C is shown in Figure 3. A clear indication of an oscillating pattern is present.

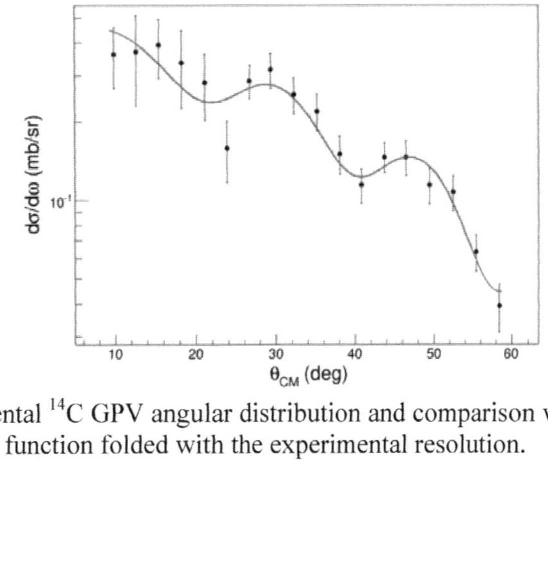

Figure 3. Experimental ^{14}C GPV angular distribution and comparison with the L = 0 Bessel function folded with the experimental resolution.

3. The GPV properties

A first important information is to identify the various components of the projectile break-up present in the inclusive excitation energy spectra, in order to isolate the spectral characteristics of the resonant-like excitations observed in the ^{15}C and ^{14}C nuclei. An extreme semi-classical model was used to describe the removal of two neutrons from the projectile and to analyse the measured energy spectra of ^{15}C in the region above S_{2n} [2], [13]. In this model, the specific treatment of nucleon-nucleon correlations beyond the residual nucleus mean field, as, for example, those due to the neutron-neutron pairing in the sd-shells were not included. The calculations give a good account for the continuum background in the excitation energy spectra. Both bumps in ^{14}C and ^{15}C are not explained within this approach. Similarly, approaches based on the towing mode of the 2n cluster [14] fail in accounting for the observed bumps [3]. These results indicate that a more complete description of the ^{12,13}C + n + n systems, including the n-n correlations, is required to describe these structures.

Another important aspect is investigation of the possible common origin of the two observed structures in ^{15}C and ^{14}C. To this aim, the excitation energy compared to the target ground state was computed as $E^t_x = E_x + M_r - M_t$, where M_t and M_r represent the mass of the target and of the residue, respectively. E^t_x is a more suitable parameter when comparing the data with theoretically derived p-p excitations built on the target mean field. When applying such a scaling of the excitation energies, the energy of the two resonances is found at about the same value in both nuclei. This is consistent with the results of ref. [15], where Quasi-particle Random Phase Approximation (QRPA) calculations predicted a GPV mode at ~ 20 MeV in oxygen isotopes. The same kind of calculations were performed for the ^{12}C response to the monopole p-p operator to predict the 0^+ two neutrons addition strength [3]. A broad distribution connected to the superposition of the $2p_{3/2}$, $2p_{1/2}$, $1d_{3/2}$, $1f_{7/2}$ and orbitals above is identified in the response function at 17 MeV with respect to the $^{14}C_{g.s}$, a value consistent with the experimental centroid of the bump. A certain degree of collectivity is observed in such a bump when the residual interaction is taken into account.

A model independent approach to investigate the multipolarity of the observed resonances is the analysis of the oscillating behaviour of the experimental angular distributions. A well-known phenomenon in heavy-ion induced transfer reactions above the Coulomb barrier is the presence of oscillations in the angular distributions only for the L = 0 transitions [16]. The distinctive feature of the L = 0 angular distributions provides a model independent identification of L = 0 among the other multipolarities. This phenomenon was experimentally observed for the same ^{12}C(^{18}O,^{16}O)^{14}C reaction at 84 MeV in ref. [1] for transitions below S_{2n}. The oscillating pattern observed in the angular distributions of the two broad resonances in ^{14}C and ^{15}C (see Figure 3 in the case of ^{14}C) supports their dominant L = 0 nature.

Another observable defining the GPV is the cross section, which was predicted to be comparable with that of the L = 0 transition to the ground state pairing vibration. In the present case, it was found $\sigma(^{14}C_{g.s.}) = 0.92$ mb and $\sigma(^{14}C_{GPV}) = 0.66$ mb integrated in the same angular range. This indicates the high-collectivity of the GPV. This property was also confirmed in terms of the strength, in order to remove trivial Q-value effects.

To summarize, the MAGNEX data showed the clearest signal compatible with the long searched GPV so far. The resonances at $E_x = 16.9 \pm 0.1$ MeV in ^{14}C and at $E_x = 13.7 \pm 0.1$ MeV in ^{15}C show properties consistent with those defining the GPV mode. These results were recently published in [3]. The study of the decaying features of these resonances, partially presented in the next section, could reveal further information on the nature of the neutron-neutron correlation in the giant mode.

4. Neutron decay spectroscopy

The investigation of the decay modes of nuclear states populated in direct reactions is a powerful tool for understanding their microscopic structure. When the reaction populates unbound or weakly bound neutron-rich systems, the neutron emission is the dominant decay mode of the excited states and sometimes of the ground state itself. In these cases, the coincidence detection of the emitted neutrons and charged ejectiles and the high-resolution measurement of the neutron energy are crucial tasks for spectroscopic investigations of the residual nuclei.

Above the one-neutron separation energy of ^{15}C, S_n = 1.218 MeV, the neutron decay of the observed resonances was studied using the EDEN array of neutron detectors [8] in coincidence with MAGNEX, as described in Ref. [12]. The technique consists in measuring the ion energy spectra (see Figure 2) using MAGNEX, then gating on the different peaks of the ^{15}C excitation energy spectrum (E_x) and plotting the corresponding neutron energy spectra (E_n) (see Figure 4) measured by EDEN in coincidence. An analysis of the neutron decay ratios was also performed. More details and the obtained results are presented in Ref. [7]. The resulting neutron emission probability for the whole ^{15}C spectrum above Sn is 101% ± 8%. Thanks to the high energy resolution in the ^{15}C spectrum, it was also possible to determine for the first time the neutron emission probability for each excited state.

Considering for example the ^{15}C excited state at E_x = 4.220 MeV, the neutron energies obtained by gating the excitation energy spectrum in the region 3.8 MeV < E_x < 4.4 MeV is shown in Figure 4a. A structure in the neutron spectrum at around 3 MeV appears well visible. Such an energy corresponds, within the uncertainty, to the energy of the neutrons decaying from the ^{15}C excited state at E_x = 4.220 MeV to the ^{14}C ground state, which is $E_n = E_x - S_n$ = 3 MeV. This means that such ^{15}C resonance decays to the ^{14}C ground state. A branching ratio of 88 ± 16% was measured. The analysis of the other states of ^{15}C below S_n is reported in ref. [7].

The neutron decay of the ^{15}C resonance at E_x = 13.7 MeV, associated to the Giant Pairing Vibration, to the ^{14}C ground state is shown in Figure 4. The measured neutron energy spectrum rules out the emission of single neutrons which are expected to have energy distributed around $E_n = E_x - S_n$ = 13.7 - 1.22 = 12.5 MeV. The high energy part of the neutron spectrum can be described as the decay to the group of ^{14}C excited states between $E'_x(^{14}C)$ = 6.094 MeV and $E''_x(^{14}C)$ = 7.341 MeV, which would produce neutrons with energies distributed between 5 and 6.4 MeV. However, the most intense neutron distribution in coincidence with the GPV peak of ^{15}C can be explained by the decay to the ^{13}C ground state via a two-neutron emission. In this case, neutron energies ranging from zero to $E_n = E_x - S_{2n}$ = 4.3 MeV are expected and, indeed, observed. Due to the low yield, two-neutron coincidences were not observed in the present experiment. A dedicated run is foreseen as the next step of this research program.

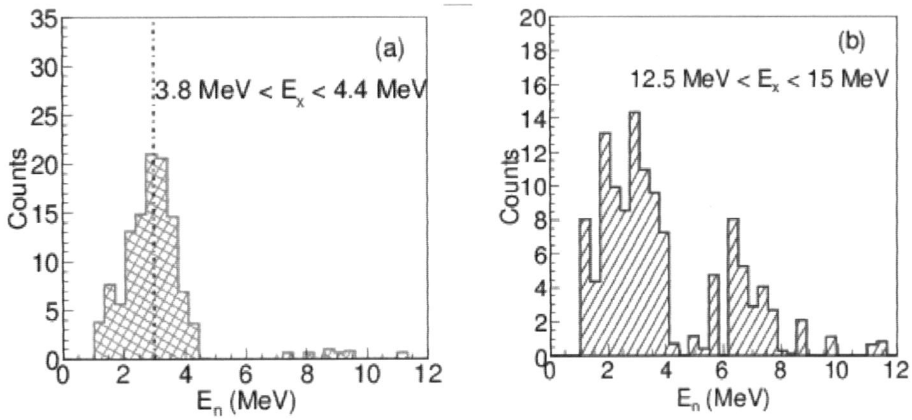

Figure 4(a) Neutron energy spectrum gated on the ^{15}C excitation energy 3.8 MeV < E_x < 4.4 MeV. (b) Neutron energy spectrum gated on the ^{15}C excitation energy 12.5 MeV < E_x < 15 MeV. In both plots the background neutron spectra are subtracted as described in ref. [7].

5. Conclusion

The Giant Pairing Vibration has been observed for the first time in the ^{14}C and ^{15}C nuclei populated via (^{18}O, ^{16}O) reactions at 84 MeV incident energy at excitation energy $E_x = 16.9 \pm 0.1$ MeV and $E_x = 13.7 \pm 0.1$ MeV, respectively. We performed a further run at 270 MeV and the results confirm the presence of the two new structures in ^{14}C and ^{15}C with the same centroid and width, within the error bars. The properties of the GPV, such as the excitation energy, the $L = 0$ angular momentum transfer, the width and the collective nature have been studied and are consistent with those defining the GPV mode. This observation confirms the basis of the particle-hole quantum symmetry in a system of interacting fermions and can provide benchmark information on the role of the pairing field in nuclear structure.

The decay by neutron emission of the states of ^{15}C above the one-neutron emission threshold and of the GPV resonance have been measured by coupling the MAGNEX magnetic spectrometer to the EDEN array of neutron detectors. The results show that the ^{15}C GPV mainly decays to the ^{13}C ground state emitting two neutrons sharing the available energy, while the decay to the ^{14}C ground state via single neutron emission is ruled out by the measured neutron energy spectrum. This is a further confirmation of the presence of large components with two correlated neutrons on a ^{13}C core for this state.

References

[1] Cavallaro M *et al.* 2013 *Phys. Rev. C* **88** 054601
[2] Cappuzzello F *et al.* 2012 *Phys. Lett. B* **711** 347
[3] Cappuzzello F *et al.* 2015 *Nat. Commun.* **6** 6743
[4] Carbone D 2015 *Eur. Phys. J. Plus* **130** 143
[5] Piekarewicz J 2015 *Nature Physics* **11** 303
[6] Broglia R A and Bes D 1977 *Phys. Lett. B* **69** 129
[7] Cavallaro M *et al.* 2016 *Phys. Rev. C* **93** 064323
[8] Laurent H *et al.* 1993 *Nucl. Instr. and Meth.* A **326** 517
[9] Cavallaro M *et al.* 2013 *Nucl. Instr. Meth. A* **700** 65
[10] Berz M *et al.* 1993 *Phys. Rev. C* **47** 537
[11] Cavallaro M *et al.* 2011 *Nucl. Instr. Meth. A* **648** 46
[12] Cavallaro M *et al.* 2011 *Nucl. Instr. Meth. A* **637** 77
[13] Carbone D *et al.* 2014 *Phys. Rev. C* **90** 064621
[14] Scarpaci J A *et al.* 1998 *Phys. Lett. B* **428** 241
[15] Khan E *et al.* 2004 *Phys. Rev. C* **69** 014314
[16] Kahana S and Baltz A J 1977 *Adv. Nucl. Phys.* **9** 1

Journal of Physics: Conference Series **730** (2016) 012008 doi:10.1088/1742-6596/730/1/012008

Few-body reaction frameworks for the study of light nuclei

R. Crespo

Instituto Superior Técnico, Universidade de Lisboa,
Av. Rovisco Pais Lisboa, Portugal

E-mail: raquel.crespo@tecnico.ulisboa.pt

Abstract. Three-body reaction formalisms, the exact Faddeev/AGS approach and the approximate reaction models such as CDCC, DWIA and Glauber multiple scattering approach are reviewed. The limits of the validity of the reaction approaches and recent developments and outlook are presented.

1. Introduction

One of the challenges of Nuclear Physics is the study of Nuclear Structure and its evolution along the nuclear landscape. Structure information from light nuclei has often been extracted from the study of its collision with light targets. It is now well established that for each target probe and for a given projectile energy, one has to identify what are the key excitation mechanisms and dynamical aspects that need to be to incorporated in order to extract the relevant structure information from the data.

We consider here that the system under study is well described by a core (either in the ground or excited state) and a valence nucleon. The collision process reduces then to a three-body problem. Special attention will be payed to the case of breakup of light halo nuclei and nucleon knockout from the collision with a proton target. In this case one expects to extract structure information from both valence and inner shells.

In this contribution we shall make a review of the available state-of-the art three-body reaction frameworks and their application for the study of nucleon knockout from a stable/exotic nucleus and breakup of a loosely bound system such as a one-nucleon halo system. We shall give insight on their physics content and revise recent benchmark calculations.

2. The Faddeev/AGS multiple scattering framework

The standard Faddeev/Alt, Grassberger and Sandhas (Faddeev/AGS) [1, 2] is a three-body nonrelativistic reaction framework.

Let us consider the few-body scattering reaction where, in inverse kinematics, the composite system, a bound pair of particles (2,3) scatters from particle 1. In our main case of interest this corresponds to the scattering of a composite system (of a core C and a valence/knockout nucleon N) from a proton target (p).

The standard approach is formulated in the Hilbert space $\mathcal{H}_{C+N+p}$ for $C+N+p$ free relative motion where the heavy fragment can only be either in the ground state or in a low-lying excited

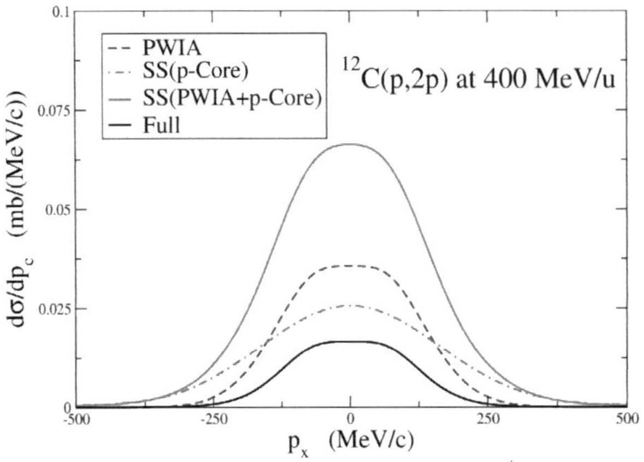

Figure 1. (Color online) Core transverse momentum distributions for ^{12}C(p,2p) at 400 MeV/u. The curves are described in the text.

state. Therefore, core dynamical excitations during the collision process, highly excited states above evaporation or multiple particle knockout (p,pNN' ..) are not taken into account in this truncated Hilbert space. This reaction formalism treats all open channels (elastic, breakup and transfer) simultaneously. In addition, all resonant and nonresonant states of each pair are automatically included.

In this approach, the transition amplitudes leading to the observables are the on-shell matrix elements of the operators $U^{\beta\alpha}$, calculated from the solution of the three-body AGS integral equations

$$U^{\beta\alpha} = \bar{\delta}_{\beta\alpha} G_0^{-1} + \sum_{\gamma} \bar{\delta}_{\beta\gamma} t_\gamma G_0 U^{\gamma\alpha} , \tag{1}$$

with $\alpha, \beta, \gamma = (1, 2, 3)$. Here, $\bar{\delta}_{\beta\alpha} = 1 - \delta_{\beta\alpha}$ and the two-body transition operator is

$$t_\gamma = v_\gamma + v_\gamma G_0 t_\gamma , \tag{2}$$

with the free resolvent $G_0 = (E + i0 - H_0)^{-1}$, with E the total energy of the three-particle system in the center of mass (c.m.) frame. The solution of the Faddeev/AGS equations at higher energies can be found by iteration

$$U^{\beta\alpha} = \bar{\delta}_{\beta\alpha} G_0^{-1} + \sum_{\gamma} \bar{\delta}_{\beta\gamma} t_\gamma \bar{\delta}_{\gamma\alpha} + \sum_{\gamma} \bar{\delta}_{\beta\gamma} t_\gamma \sum_{\xi} G_0 \bar{\delta}_{\gamma\xi} t_\xi \bar{\delta}_{\xi\alpha} + \cdots . \tag{3}$$

The successive terms of this series can be considered as terms of zero order (which contribute only for rearrangement transitions), first order (single scattering), second order (double scattering) and so on in the transition operators t_γ. The $\beta = 0$ partition corresponds to three free particles in the continuum.

The breakup/nucleon knockout observables are calculated from the on-shell matrix elements of the AGS operators, $T^{0\alpha} = \langle \mathbf{qp} | U^{0\alpha} | \psi_\alpha \rangle$, where particle α is the spectator in the initial state (in our case α is the proton target) and $\mathbf{qp}$ are the Jacobi momenta in the final state

$$\begin{aligned} \mathbf{p} &= \frac{m_N \mathbf{K}_p - m_p \mathbf{K}_N}{m_p + m_N} , \\ \mathbf{q} &= \frac{(m_p + m_N)\mathbf{K}_C - m_C(\mathbf{K}_p + \mathbf{K}_N)}{M} , \end{aligned} \tag{4}$$

with $\mathbf{K}_p$, $\mathbf{K}_N$, $\mathbf{K}_C$ (m_p, m_N, m_C) being the LAB momenta (masses) of the proton, valence/knockout nucleon and core in the exit system, and $M = m_p + m_N + m_C$. The kinematically fully exclusive observables are measured in the LAB system. The kinematic configuration of three-body final system is characterized by the polar and azimuthal angles $\Omega_i = (\theta_i, \phi_i)$ of the two detected particles which we take to be the valence/knockout nucleon and the proton target p. The kinematically fully exclusive cross section is then

$$\frac{d^5\sigma}{d\hat{\mathbf{K}}_N d\mathbf{K}_p} = \sum_i \left[\, |T^{0\alpha}|^2 \rho_i \,\right]_i , \tag{5}$$

where ρ_i is a phase-space factor given by

$$\rho_i = (2\pi)^4 \frac{(m_N + m_p) m_C m_N}{K_{\rm LAB}} \frac{K_n^2}{|(m_N + m_C) K_N - m_N (\mathbf{K}_{\rm LAB} - \mathbf{K}_p) \cdot \hat{\mathbf{K}}_N|} , \tag{6}$$

and the sum on i involves the momenta K_N given by the zero's of the argument of the energy conserving δ-function

$$E_{\rm LAB} - \epsilon - \frac{(\mathbf{K}_{\rm LAB} - \mathbf{K}_N - \mathbf{K}_p)^2}{2m_C} - \frac{K_N^2}{2m_N} - \frac{K_p^2}{2m_p} = 0 , \tag{7}$$

where $\epsilon = E^* - Q_0$ being E^* the excitation energy of the heavy fragment, Q_0 the mass excess, $\mathbf{K}_{\rm LAB}$ is the beam momentum in the LAB frame and $E_{\rm LAB}$ the corresponding energy.

In the standard Faddeev/AGS reaction framework, the single scattering amplitude, includes the term where the proton target scatters from the knockout nucleon (p–N) and from the heavy fragment (p–C). As one can see from the Fig. 1, the single scattering contribution where the proton target scatters from the heavy fragment (dashed-dotted line) is very significant. The case where one includes only the single scattering of the knockout nucleon by the proton target is standardly referred as the Plane Wave Impulse Approximation (PWIA). Within the Faddeev/AGS reaction framework, distortion effects with respect to the PWIA approximation result from the simultaneous combination of the p–C single scattering term and of higher order multiple scattering contributions. From Eq. 5, it can be seen that formally the distortion effects do not factorize into a renormalization factor to the PWIA term. Nevertheless, a systematic study made in [3] has shown that, at high energies, there are subtle cancellations between the single scattering (p–C) and higher order multiple scattering terms such that (under the assumption of a single core state) a distortion parameter (defined as the relative difference between the PWIA and the full Faddeev/AGS total cross section) has a nearly logarithmic dependence on the separation energy of the knockout nucleon, for all considered values of the angular momentum of the knockout particle. A timely result was also given in Ref. [3]: the calculated ratio of the full to the plane wave total cross section was found to have a roughly linear trend when expressed as a function of the asymmetry parameter with a smaller slope than the ratio of the theoretical to the experimental total cross section for nucleon knockout found in the work of [15]. Further insight on the dependence of the distortion with respect to the angle of the light particles will be of interest for future experiments at Radioactive Ion Beam Facilities.

One key aspect that needs to be addressed is the validity of the choice of the truncated Hilbert space at a given energy regime. Standard developments of the Faddeev/AGS reaction approach assume a truncated Hilbert space for the three particle system, in particular it presupposes an inert core configuration. Recently, the Faddeev/AGS momentum-space framework for nuclear reactions has been extended to include also excitations of the core nucleus. In this case, the contribution of the transition amplitude for the scattering between the Core and the proton

Journal of Physics: Conference Series **730** (2016) 012008 doi:10.1088/1742-6596/730/1/012008

target induces the core excitation mechanism. Important core excitation effects were found at intermediate energies for the breakup of ^{11}Be on a proton target at 64 MeV/u [4] and at low energies for the case of the transfer reaction $^{10}\mathrm{Be}(d,p)^{11}\mathrm{Be}$ at $E_d = 15$ and 21.4 MeV leading to the ground state $1/2^+$ of ^{11}Be [5].

3. The DWIA

The distorted-wave impulse-approximation (DWIA) reaction framework [6], can be viewed as an incomplete and truncated multiple scattering series [7], commonly used to analyze the knockout reaction $A(a,ab)B$ where an incident particle a knocks out a nucleon or a bound cluster b in the target nucleus A resulting in three particles (a,b,B) in the final state.

This reaction framework relies mainly in two basic assumptions: (i) that the projectile struck the ejected particle freely and (ii) that the 3-body wave function for the final state can be written as a factorized product of two wave functions that describe the $a+B$ and $b+B$ scattering $\eta^{(-)}_{Bab} \sim \eta^{(+)}_{aB}\eta^{(+)}_{bB}$. The transition amplitude is written as

$$T_{AB} = \langle \eta^{(-)}_{aB}\eta^{(-)}_{bB}|t_{ab}|\phi_{Bb}\eta^{(+)}_{aA}\rangle, \tag{8}$$

where the wave function for the relative motion of the particles in the entrance channel satisfy

$$(T_{aA} + V_{aA} - V_{ab} - \epsilon_{aA})\eta^{(+)}_{aA} = 0, \tag{9}$$

where ϵ_{aA} is the relative kinetic energy and T_{aA} the relative kinetic energy operator. For the exit channel, the relative wave functions satisfy:

$$(T_{aB} + V_{aB} - \epsilon_{aB})\eta^{(+)}_{aB} = 0, \tag{10}$$

and

$$(T_{bB} + V_{bB} - \epsilon_{bB})\eta^{(+)}_{bB} = 0. \tag{11}$$

The potentials V_{aB} and V_{bB} are taken to be the optical potentials which describe the $a+B$ and $b+B$ scattering at energies ϵ_{aB} and ϵ_{bB} respectively. Further approximations are usually made in standard applications of the DWIA when evaluating the transition amplitude, namely (i) the neglect of the spin-orbit interactions in the distorting potentials (ii) the factorization approximation, which is only exact in PWIA (iii) the on-shell approximation of the transition amplitude. For (p,pN) reactions this leads to the the result

$$\frac{d^5\sigma}{d\hat{\mathbf{K}}_N d\mathbf{K}_p} = \mathcal{F}\sum_{\lambda} |\mathcal{P}^{\lambda}(\mathbf{K}_C, E)|^2 \frac{d\sigma}{d\Omega}_{\mathrm{pN}}(E_{\mathrm{LAB}}, \mathbf{K}_p, \hat{\mathbf{K}}_N) \tag{12}$$

where $\mathcal{F}$ is a kinematic factor, λ the relevant quantum numbers and the quantity $\sum_{\alpha} |\mathcal{P}^{\lambda}(\mathbf{K}_C, E)|^2$ represents the distorted momentum distribution of the knockout particle in the composite system. We note that extracting a spectroscopic factor from comparing with the data the calculated cross section using the DWIA framework relies on the validity of writing the transition amplitude as a product of a structure factor with a dynamical cross section (for the scattering between the proton and the knockout particle). This is an approximation, that needs to be tested and validated as discussed in detail in the previous section. In addition, one should also note that the distortion momentum distribution can only be uniquely determined from Eq. 12 if the ratio of the knockout cross section to the p-N cross section is independent of the p-N angle pair. Recent applications of the DWIA approach make further use of eikonal-forward approximation in the evaluation of the distorted momentum distributions [8, 9]. The eikonal approximation produces an additional truncation of the incomplete multiple scattering expansion to second order.

Journal of Physics: Conference Series **730** (2016) 012008 doi:10.1088/1742-6596/730/1/012008

4. The continuum discretization reaction approach

The continuum discretized reaction method consists of an approximation to the Faddeev formalism. It makes use of a truncated model space where the three-body wave function is expanded in terms of all states of a given interacting pair.

In this reaction formalism, assuming that the projectile is well described by an inert core C and a valence particle v, the Schrödinger equation is commonly solved in a model space in which, the three-body wave function is expanded in the internal states (bound and continuum resonant and nonresonant states) of the two-body projectile [10]. The exact three-body wave function satisfies the Schrödinger equation

$$(H - E)\Psi^{+}_{\mathbf{K}_0}(\mathbf{R}, \mathbf{r}) = 0 \ , \tag{13}$$

where H is the three-body Hamiltonian, E the total c.m. energy of the system, $\mathbf{R}$ the relative distance between the c.m. of the projectile and the target and $\mathbf{r}$ the relative distance between the valence particle and the core. Finally $\mathbf{K}_0$ is the incident wave number of the projectile in the c.m. frame.

The Hamiltonian H for this system can be written as a sum of the relative motion between the projectile and target and the internal Hamiltonian of the projectile $H_{\rm int}$, $H = T_R + \sum_{j=C,v} V_{jt}(\mathbf{R}, \mathbf{r}) + H_{\rm int}$ where T_R is the c.m. kinetic energy. The exact wave function is then expressed as an expansion in terms of the states (bound and bin) $\phi_\alpha(\mathbf{r})$ of the two-body Hamiltonian $H_{\rm int}$

$$\Psi^{\rm CDCC}_{\mathbf{K}_0}(\mathbf{R}, \mathbf{r}) = \sum_{\alpha=0}^{N} \phi_\alpha(\mathbf{r})\omega_\alpha(\mathbf{R}) \tag{14}$$

where α=0 refers to the projectile ground state. The bin states include both the resonant and the nonresonant part of the continuum. The wave functions $\omega_\alpha(\mathbf{R})$ of the projectile-target relative motion are solutions of the coupled-channels equations

$$\left(E_\alpha - T_R - V_{\alpha\alpha}(\mathbf{R})\right)\omega_\alpha(\mathbf{R}) = \sum_{\beta\neq\alpha} V_{\alpha\beta}(\mathbf{R})\omega_\beta(\mathbf{R}) \ , \tag{15}$$

where $E_\alpha = E - \varepsilon_\alpha$ and $V_{\alpha\beta}(\mathbf{R}) = \langle\phi_\alpha|\sum_{j=C,v} V_{jt}(\mathbf{R}, \mathbf{r})|\phi_\beta\rangle$ the coupling potentials. Alternatively for breakup/nucleon knockout the three-body wave function can be expanded in the p-N states [11, 12].

This formalism was extended recently to include core dynamical excitations [13, 14].

5. The adiabatic-eikonal Glauber formalism

The adiabatic-eikonal Glauber approach can be understood in terms of the multiple scattering Watson series [7] and makes use of two key approximations (i) the eikonal forward scattering approximation leading to a second order truncation of the series (ii) the adiabatic or sudden approximation which consists in replacing the internal hamiltonian of the few-body system by a constant $\bar{H}$. In this framework, the projectile resolvent $\hat{G}(z) = (z - H_0 - \sum V_{\mathcal{I}\mathcal{J}})^{-1}$ where $V_{\mathcal{I}\mathcal{J}}$ is the interaction between subsystems $\mathcal{I}$ and $\mathcal{J}$ of the composite projectile is replaced by

$$\hat{G}(z) \rightarrow G^{\rm eik} = \left(z - K_P - \bar{H}\right)^{-1} \tag{16}$$

in the transition amplitude for the composite projectile cluster - target pair

$$\tau^{\rm eik}_{\mathcal{I}} = v_{\mathcal{I}} + v_{\mathcal{I}}\hat{G}^{\rm eik}\tau^{\rm eik}_{\mathcal{I}} \ . \tag{17}$$

This means that rescattering terms between the subsystem of the composite projectile are taken into account in an approximate way and the validity of this approximation needs to be assessed if one wants to extract accurate structure information from the data [7, 19].

Applications of this framework have been done for the analysis of inclusive target- and core- inclusive knockout data for light nuclei [15] (and references therein). In these analysis completeness of the bound and continuum two-body eigenstates of the composite projectile is used and it is further assumed that contributions from single particle excitations can be neglected [16].

6. Reaction formalisms validity barometer

To have a reliable interpretation of reaction data, as a first step one needs to include in the reaction formalism all the relevant excitation mechanisms, that is one has to include all the relevant degrees of freedom in the Hilbert space for the projectile-target system.

In addition to this, one also needs to include accurately all the dynamics of the reaction. Reaction methods often include approximations to the treatment of the dynamics. The validity of these approximations needs to be estimated.

Benchmark calculations for elastic, breakup, and transfer observables using the continuum discretized and the Faddeev/AGS standard reaction frameworks [11, 17] exhibit some discrepancies. The results indicate that for the case of elastic scattering and breakup of deuterons on ^{12}C at E_d=56 MeV the continuum discretized is in agreement with the AGS/Faddeev three-body results. However, for deuteron breakup a large disagreement is found at the lower energy of E_d=12 MeV. For the p-^{11}Be test case, while for elastic scattering at 39 MeV/u the two methods are in good agreement, the calculated transfer cross section at 38.4 MeV/u using the continuum discretized reaction framework underestimates the result obtained from solving the Faddeev/AGS equations. For the breakup observables, good agreement between the two reaction approaches can be obtained in certain regions of the phase space, depending on the choice of basis used for the the expansion of the wave function. It was also found that in the case of the transfer cross section for deuterons on ^{10}Be the disagreement between the two reaction frameworks increases with energy.

In the work of Ref. [7] an attempt was made to bridge the exact standard Faddeev/AGS reaction framework with the approximate Glauber formalism for the case of elastic scattering. It was shown in this work that third order rescattering terms absent from the Glauber reaction formalism due to the eikonal approximation are important to describe elastic scattering. Further insight on the validity of the adiabatic approximation used in Glauber calculations is necessary, therefore benchmark calculations for other reaction channels need to be performed in the future.

In the work of Refs. [18, 19] an attempt was also made to bridge the exact standard Faddeev/AGS reaction framework with the approximate DWIA formalism. In that work it was shown that the DWIA transition amplitude can be approximately expressed in terms of a multiple scattering (MS) expansion series referred as DWIA-MS, and shown therein to be incomplete. The DWIA-MS containing multiple scattering terms up to second order, underestimate the corresponding second order Faddeev/AGS result by about 20%. The agreement found between the 2nd order DWIA-MS results and the converged Faddeev/AGS result, is unclear. Further work is needed to understand if there are cancellations between higher order multiple scattering terms or if this agreement is furtuitos.

Bechmark calculations between the two reaction frameworks are also needed in the future.

7. Conclusions and outlook

We have reviewed the theory of the standard Faddeev/AGS few-body reaction framework along with other standard formalisms (DWIA, CDCC and Glauber). When viewed as a multiple scattering expansion in terms of transition operators for the interacting pairs, these approximate reaction frameworks are described by truncated and incomplete multiple scattering series. In the Faddeev/AGS framework the sum of the series can be included. Benchmark calculations are needed to estimate the validity of approximate treatments of the dynamics of the reaction.

In addition, for the energy regime under consideration, the contribution of other excitation mechanisms such as core dynamical excitations during the collision process, highly excited states above evaporation or multiple particle knockout (p,pNN' ..) should also be estimated.

Acknowledgments
The author would like to thank E. Cravo and A. Deltuva for reading this manuscript.

References

[1] Faddeev L D 1960 *Zh. Eksp. Teor. Fiz.* **39** 1459 [Sov. Phys. JETP **12**, 1014 (1961)]
[2] Alt E O, Grassberger P and Sandhas W 1967 *Nucl. Phys.* **B2** 167
[3] Cravo E, Crespo R, Deltuva A, submitted for publication.
[4] Crespo R, Deltuva A and Moro A M 2011 *Phys. Rev. C* **83** 044622
[5] Deltuva A 2013 *Phys. Rev. C* **88** 011601(R)
[6] Chant N and Roos P G 1977 *Phys. Rev. C* **15** 57
[7] Crespo R, Cravo E, Deltuva A, Rodriguez-Gallardo M and Fonseca A C 2007 *Phys. Rev. C* **76** 014620
[8] Aumann T A, Bertulani C A, and Ryckebusch J, 2013 2015 *Phys. Rev. C* **88**, 064610
[9] Ogata K, Yoshida K, Minomo K, 2015 *Phys. Rev. C* **92**, 034616
[10] Austern N, Iseri Y, Kamimura M, Kawai M, Rawitscher G and Yahiro M 1987 *Phys. Rep.* **154** 125
[11] Deltuva A, Moro A M, Cravo E, Nunes F M and Fonseca A C 2007 *Phys. Rev. C* **76** 064602
[12] Moro A M 2015 *Phys. Rev. C* **92**, 044605
[13] Summers N C, Nunes F M and Thompson I J, 2006 *Phys. Rev. C* **74**, 014606
[14] Diego R, Arias J M, Lay L A, Moro A M, 2014 *Phys. Rev. C* **89**, 064609
[15] Tostevin J A and Gade A, 2014 *Phys. Rev. C* **90**, 057602
[16] Hansen P G and Tostevin J A, 2003*Ann. Rev. Nucl. Part. Sci* **53**, 219
[17] Upadhyay N J, Deltuva A and Nunes F M 2012 *Phys. Rev. C* **85** 054621
[18] Crespo R, Deltuva A, Cravo E, Rodríguez-Gallardo M, and . Fonseca A C, 2008 *Phys. Rev. C* **77**, 024601
[19] Crespo R, Deltuva A, Cravo E, 2014 *Phys. Rev. C* **90**, 044606

Journal of Physics: Conference Series **730** (2016) 012009 doi:10.1088/1742-6596/730/1/012009

Color reconnection: a fundamental ingredient of the hadronisation in p-p collisions

E. Cuautle, S. Iga, A. Ortiz and G. Paić

Instituto de Ciencias Nucleares, Universidad Nacional Autónoma de México, A.P. 70 - 543, 04510 Ciudad de México, México

E-mail: ecuautle@nucleares.unam.mx, sergio.arturo.iga.buitron@cern.ch, antonio.ortiz.velasquez@cern.ch, guy.paic@cern.ch

Abstract. At the LHC very interesting similarities among different colliding systems (p-p, p-Pb and Pb-Pb) where observed in the multiplicity evolution of the transverse momentum spectra. This has prompted a number of analyses that have explained the results in terms of collective hydrodynamic flow. The explanation in terms of hydrodynamics has recognized problems with the smallness of the interaction volume in the systems created in p-p and p-Pb collisions. On the other hand, some event generators based on QCD produce a reasonable qualitative, and sometimes quantitative, agreement with the data. Those results can be achieved introducing in the hadronisation model the so-called color reconnection which produces flow-like patterns via boosted strings. In this work we present the behavior of the various color reconnection (CR) schemes compared to those without the CR case for different center-of-mass energies at the LHC (0.9, 7 and 13 TeV).

1. Introduction

Heavy-ion collisions at ultra-relativistic energies are used to study the matter at high energy density, where the formation of a strongly-coupled Quark-Gluon-Plasma (sQGP) has been established. The discovery at the LHC is that in small systems, like those created in p-p and p-Pb collisions, collective features have been found [1]. It has been shown that the QCD based color reconnection may induce collective-like effects. In this work we demonstrate the influence of different CR models on the observed transverse momentum spectra.

There is evidence that in p-p collisions multiple partonic interactions (MPI) may occur within the same collision[2]. The partons belonging to two different MPI can be reconnected by color strings and merged into one in such a way that the total string length becomes as short as possible.

There are three different CR models included in PYTHIA 8.210 [3]

- **MPI-based model:** The partons of a lower-p_T MPI system are merged with the ones in a higher-p_T MPI.
- **New scheme:** The QCD color rules are used to determine the reconnection probability.
- **Gluon motion:** Partons can be moved from one location to another so as to reduce the total string length.

Journal of Physics: Conference Series **730** (2016) 012009 doi:10.1088/1742-6596/730/1/012009

These models are used to generate the transverse momentum spectra to compare the collective-like effects previously studied[5].

2. Results and discussion

For this analysis 100 million events were generated for each simulation(three CR models plus the no CR case) of inelastic p-p collisions at $\sqrt{s} = 0.9$, 7 and 13 TeV. For the particle selection we only take the charged particles within $|\eta| < 0.8$, removing products of weak decays. All the PYTHIA parameters were set to Monash Tune[4], except for the CR model.

In figure 1 we show the multiplicity distributions for different energies and different CR models. There are little differences between the three CR models if we look at the same $\sqrt{s}$ but there are notorious differences from the no CR case, it increase as the energy do. This behaviour is explained in the way that with CR we have the merging of different MPI's and this reconnection produces lesser number of particles[6].

After the low multiplicity part dominated by diffractive events we observes a clear indication of the increasing importance of the high multiplicity events in the total multiplicity.

To quantify the different CR models we define multiplicity classes $z = N_{ch}/\langle N_{ch} \rangle$ for the p_T spectra. In figure 2 the CR-to-noCR ratio as function of p_T is presented for two multiplicity classes for p-p collisions at $\sqrt{s} = 7$ TeV, the spectra used for the ratio are normalized to the number of events in each multiplicity region. We observe that the effects of CR are stronger for the high multiplicity events (right panel). The CR at high multiplicity depletes the low

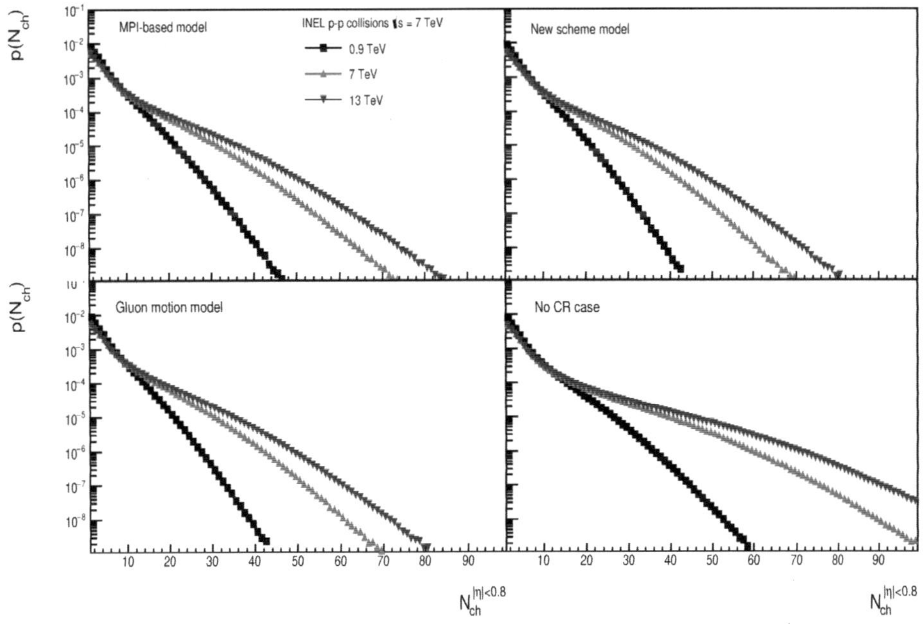

Figure 1. Multiplicity distributions for different energies and CR models: MPI-based (top left), new scheme (top right), gluon motion (bottom left) and without CR (bottom right).

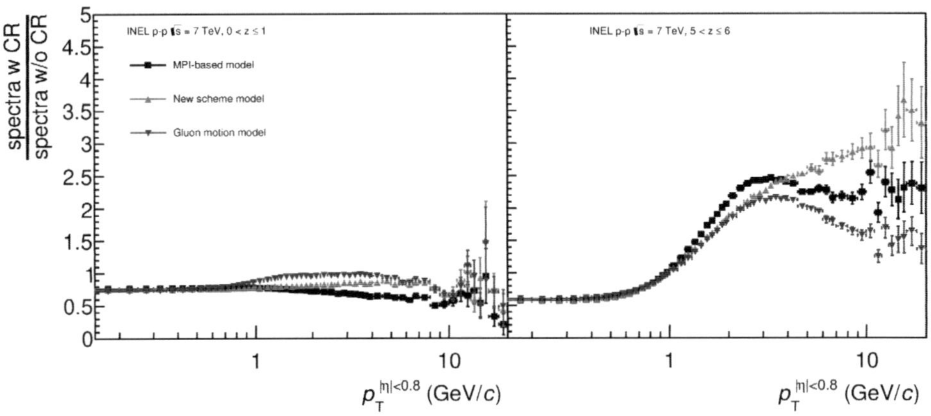

Figure 2. CR-to-noCR ratios of the inclusive p_T spectra for different CR models at $\sqrt{s} = 7$ TeV for low multiplicity class $0 < z \leq 0.5$ (left) and high multiplicity class $5 < z \leq 6$ (right).

momentum region and enhance the high momentum one. This ratio trends to a constant for the low multiplicity region (figure 2 left panel) and goes closer to one if we normalize the spectra for the ratios to the area instead of the number of particles[7], suggesting that there are no CR effects on events of low multiplicity (lower number of MPI's). The same trends is observed for other energies.

In figure 3 we show the p_T distribution of the particle with the largest p_T of the event (leading p_T), the spectra are normalized to the number of events in each multiplicity interval. The leading p_T distributions are presented for events with different multiplicities and different CR cases. We see that for high multiplicity events a shift of the maximum of the distribution at higher p_T values occurs for the CR case (MPI-based model) with respect to the no CR one, we see this shift too in the other two CR models (new scheme and gluons motion).

In table 1 we show the mean leading p_T values for three multiplicity intervals and for the CR MPI-based model compared to the no CR case in function of the center of mass energy. An increase with the multiplicity is observed, the effect is bigger with CR.

In figure 4 we show the CR-to-noCR ratio for the leading p_T spectra showed in figure 3.

	$\sqrt{s} = 0.9$ TeV		$\sqrt{s} = 7$ TeV		$\sqrt{s} = 13$ TeV	
z region	CR, MPI-based	No CR	CR, MPI-based	No CR	CR, MPI-based	No CR
$0 < z \leq 1$	0.71	0.71	0.85	0.85	0.89	0.90
$3 < z \leq 4$	1.81	1.53	2.73	2.19	3.00	2.39
$z > 5$	2.37	1.87	3.69	2.82	4.04	3.07

Table 1. Mean transverse momentum of the leading particle in z classes for different energies, comparison between CR MPI-based model and no CR.

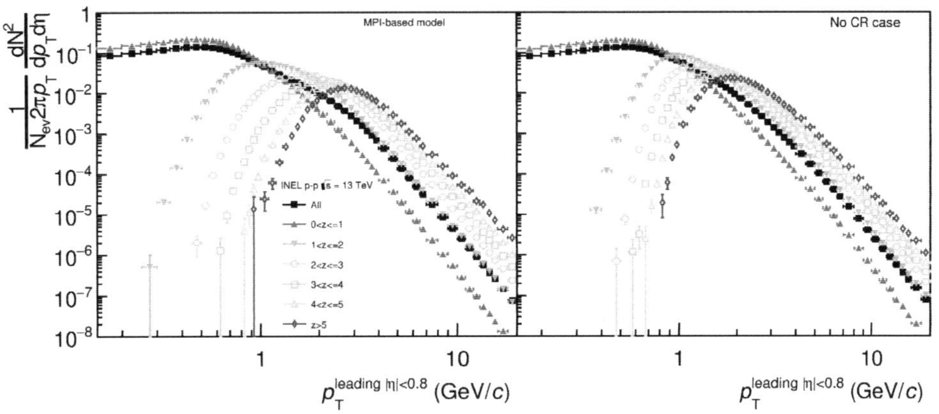

Figure 3. p_T leading distribution for different multiplicity classes at $\sqrt{s}$ = 13 TeV for CR MPI-based model (left) and no CR (right).

Again we see that also for the leading particles the effects of the CR are very small for the lowest multiplicity class, showed in in Fig.3 for the case $0 < z < 1$. For the high multiplicity classes the main difference with the inclusive case (shown in figure 2) are at low p_T values, this shift effect is more notorious in the leading case.

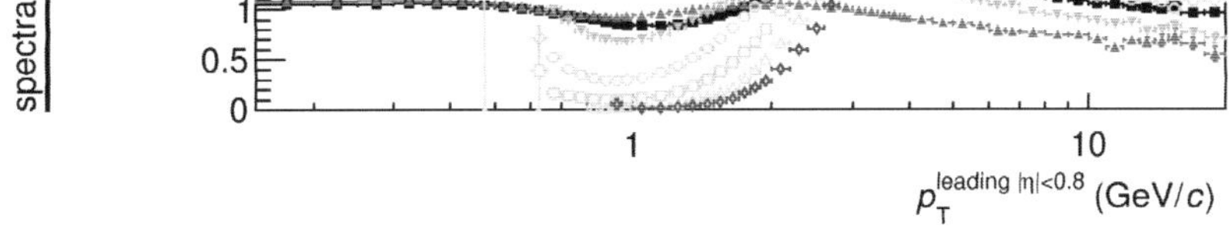

Figure 4. CR-to-noCR ratios of the leading p_T spectra in z classes for CR MPI-based model.

3. Conclusions

In the present work we demonstrated the effects of different color reconnection models to the p_T and leading p_T spectra in p-p collisions. The effects seems to have an increasing impact up to rather high momenta simulated in the present work increasing with the multiplicity of events. At very low multiplicity, events with less than half the mean multiplicity the effect of color reconnection is very weak, suggesting that the behaviour of the reconnection in experiments could be studied with respect to the lowest multiplicity region.

Acknowledgments

Support for this work has been received from CONACYT under the grant No. 270067; from DGAPA-UNAM under PAPIIT grants IN108414.

References

[1] Loizides C 2016 *Experimental overview on small colliding systems at the LHC* arXiv:1602.09138[nucl-ex]
[2] Chekanov S *et al.* (ZEUS Collaboration) 2008 *Nucl. Phys. B* **792** 1
Abe F *et al.* (CDF Collaboration) 1997 *Phys. Rev. Lett.* **79** 584
Abazov V *et al.* (D0 Collaboration) 2010 *Phys. Rev. D* **81** 052012
[3] Sjstrand T *et al.* 2015 *Comput. Phys. Commun.* **191** 159
[4] Skands P, Carrazza S, and Rojo J 2014 *Tuning PYTHIA 8.1: the Monash 2013 tune* arXiv:1404.5630
[5] Ortiz A, Christiansen P, Cuautle E, Maldonado I, and G. Paić 2013 *Phys. Rev. Lett.* **111** 042001
[6] Christiansen J R and Skands P Z 2015 *JHEP* **08** 003
[7] Cuautle E, Iga S and Paić G 2015 *Leading particle spectra as a tool to tune the Color reconnecion models* to be published as proceeding of the *7th International Workshop on Multiple Partonic Interactions at the LHC* Trieste Italy, November 23-27, 2015

Amplitude analysis of $\bar{K}N$ scattering

César Fernández-Ramírez

Instituto de Ciencias Nucleares, Universidad Nacional Autónoma de México, A.P. 70-543, México D.F. 04510, México

E-mail: cesar.fernandez@nucleares.unam.mx

Abstract. We present the results of a coupled-channel model for $\bar{K}N$ scattering in the resonance region. The model fulfills unitarity, has the correct analytical properties for the amplitudes and the partial waves have the right threshold behavior. The parameters of the model have been established by fitting single-energy partial waves up to $J = 7/2$ and up to 2.15 GeV of center-of-mass energy. The Λ^* and Σ^* spectra has been obtained, providing a comprehensive picture of the $S = -1$ hyperon spectrum. We use the structure of the hyperon spectrum and Regge phenomenology to gain insight on the nature of the $\Lambda(1405)$ resonances.

1. Introduction

The comprehensive understanding of the strong interaction in the hadronic energy range is one of the open problems in particle and nuclear physics. Problems such as non-perturbative aspects of Quantum Chromo Dynamics (QCD), how quarks and gluons aggregate to constitute hadrons, confinement and chiral symmetry breaking have spectroscopy of excited baryons constituted by light valence quarks (u, d, and s) as one of their primary tools for research. For these reasons, many experiments have measured $\pi^{\pm}N$ and $K^{-}p$ scattering as well as photoproduction off the proton in order to garner information on the spectrum. However, the amount of experimental information to pin down hyperon resonances with $S = -1$ ($Y^* = \Lambda^*, \Sigma^*$) is not as wealthy as for the case of nucleon excitations ($S = 0$) and, as a consequence, its spectrum is less understood. Only in recent years we are starting to have a good grasp of the Y^* spectrum and pole positions have started to be reported by the *Review of Particle Physics* (RPP) [1] thanks to the development of models for kaon electroproduction [2] and $\bar{K}N$ scattering [3, 4]. Once a baryon resonance and its location in the complex plane have been well-established, it is time to understand if the nature of that state is (mostly) that of a three quark system, a molecule or a pentaquark, and to find out what is the role of gluonic excitations for that particular state. In section 2 we provide a brief description of the amplitude analysis of $\bar{K}N$ scattering performed in [3] that allowed to provide the most comprehensive picture of the $S = -1$ hyperon spectrum to the date. In section 3 we use the structure of the hyperon spectrum and Regge phenomenology to gain insight on the nature of the $\Lambda(1405)$ resonance, the first excitation of the isospin-0 *uds* system and a long-standing problem in hadron spectroscopy.

2. $\bar{K}N$ scattering in the resonance region

The $\bar{K}N \to \bar{K}N$ reaction, besides its importance for studies of the Y^* spectrum, plays a role in amplitude analysis of more complicated reactions, which include decays (pentaquark searches) [5, 6] and $K\bar{K}$ pair photoproduction [7, 8]. For example, the recent observation of two pentaquark states in $\Lambda_b^0 \to J/\psi\, K^- p$ decay [5] uses a specific model to incorporate Y^* resonances in the $K^- p$ channel. Studies of systematic uncertainties should involve the comparison with other models of $\bar{K}N$ interactions and should incorporate background effects, not just Y^* excitations [6]. Real and quasi-real diffractive photoproduction of $K\bar{K}$ pairs can produce the poorly known [7], *i.e.*, mesons containing $s\bar{s}$ pairs that also include exotic mesons with hidden strangeness. The factorization of the $K\bar{K}$ photoproduction vertex requires separation of target fragmentation at the amplitude level. Hence, the provision of amplitudes describing the $\bar{K}N$ interactions in target fragmentation is relevant to future partial-wave analyses of the $\gamma p \to K\bar{K}p$ process. At Jefferson Lab [8], both CLAS12 (Hall B) and GluEx (Hall D) experiments will devote part of their effort to study this reaction. Hence, it is timely to develop a model for $\bar{K}N \to \bar{K}N$ scattering that can be incorporated in the analysis of both three-body decay and two kaon photoproduction experiments.

In [3] we presented a coupled-channel model for $\bar{K}N \to \bar{K}N$ scattering in the resonance region that incorporates up to 13 channels per partial wave, analyticity, unitarity, and the right angular momentum barrier for the partial waves. All the details can be found in [3] and here we sketch the building blocks of the model.

The partial-wave expanded S_ℓ matrix is related to the amplitude T_ℓ through

$$S_\ell = \mathbb{I} + 2iR_\ell(s) = \mathbb{I} + 2i\,[C_\ell(s)]^{1/2}\,T_\ell(s)\,[C_\ell(s)]^{1/2}\,, \tag{1}$$

where $\mathbb{I}$ is the identity matrix and the diagonal matrix

$$C_\ell(s) = \frac{q_k(s)}{q_0}\left[\frac{r^2 q_k^2(s)}{1 + r^2 q_k^2(s)}\right]^\ell\,, \tag{2}$$

accounts for the phase space, where $q_k(s) = \sqrt{(m_1 m_2)(s - s_k)}/(m_1 + m_2)$, s_k is the threshold center-of-mass energy squared of the corresponding channel k, m_1 and m_2 are the masses of the final states, $q_0 = 2$ GeV is a normalization factor for the momentum in the resonance region, $r = 1$ fm is an effective interaction range parameter and $T_\ell(s)$ is the analytical partial-wave amplitude matrix. We write $T_\ell(s)$ in terms of a K matrix [9] to ensure unitarity

$$T_\ell(s) = \left[K(s)^{-1} - i\rho(s,\ell)\,\right]^{-1}\,, \tag{3}$$

where $\rho(s,\ell)$ is the dispersive integral over the phase space matrix $C_\ell(s)$,

$$i\rho(s,\ell) = \frac{s - s_k}{\pi}\int_{s_k}^{\infty}\frac{C_\ell(s')}{s' - s}\frac{ds'}{s' - s_k}. \tag{4}$$

The $K(s)$ matrix in equation (3) is built as the addition of up to six K matrices

$$[K(s)]_{kj} = \sum_a x_k^a\, K_a(s)\, x_j^a\,. \tag{5}$$

Each one of the $K_a(s)$ matrices can be either a *pole* K-matrix

$$[K_P(s)]_{kj} = x_k^P\,\frac{M_P}{M_P^2 - s}\,x_j^P\,, \tag{6}$$

or a *background* K matrix

$$[K_B(s)]_{kj} = x_k^B\,\frac{M_B}{M_B^2 + s}\,x_j^B\,, \tag{7}$$

depending on the considered partial wave. The free parameters x_j^P, x_j^B, M_B and M_P will be fixed by fitting single-energy partial waves from [10].

In [10] the experimental database in the resonance region with $2.19 < s < 4.70$ GeV^2 (approximately 17500 data points for the $K^-p \to K^-p$, $K^-p \to \bar{K}^0n$, $K^-p \to \pi^0\Lambda$, $K^-p \to \pi^0\Sigma^0$, $K^-p \to \pi^-\Sigma^+$, and $K^-p \to \pi^+\Sigma^-$ channels) was analyzed and the single-energy partial waves were obtained for $(\ell_{I\,2J})$ up to $J = 7/2$, *i.e.* S_{01}, P_{01}, P_{03}, D_{03}, D_{05}, F_{05}, F_{07}, G_{07}, S_{11}, P_{11}, P_{13}, D_{13}, D_{15}, F_{15}, F_{17}, and G_{17}. We fitted our model to these single-energy partial waves employing MINUIT [11] and a genetic algorithm [12]. The uncertainties were computed using the bootstrap technique [13]. The value of all parameters and the codes to compute the partial waves and the observables are available from the Joint Physics Analysis Center webpage [14, 15]. Once the parameters of the model have been established we can analytically continue the partial waves to the unphysical Riemann sheets and search for poles that correspond to the hyperon resonances. Figures 1 and 2 show the Λ^* and Σ^* resonances found in the different partial waves.

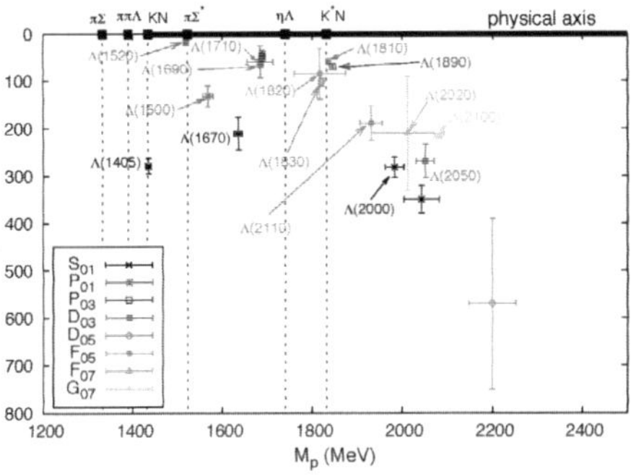

Figure 1. Isospin 0 (Λ^*) poles. Vertical dashed lines mark the different production thresholds.

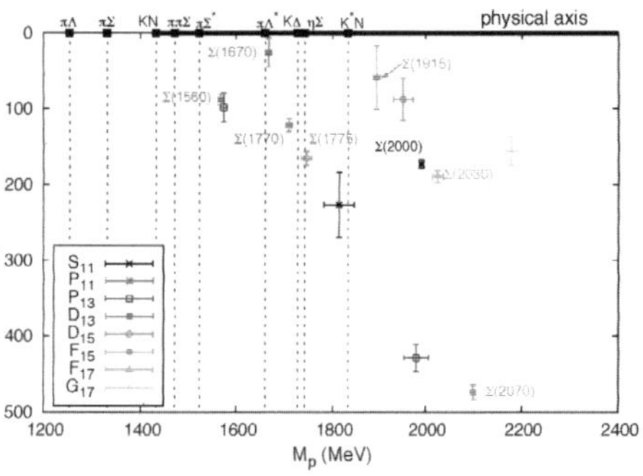

Figure 2. Isospin 1 (Σ^*) poles. Vertical dashed lines mark the different production thresholds.

3. The nature of the $\Lambda(1405)$ resonance(s)

Since its discovery in the early 60's [16], the nature of the $\Lambda(1405)$, the first excitation of the isospin-0 *uds* system, is a long-standing problem in hadron spectroscopy. It has regained interest due to the recent experimental confirmation of its, for decades assumed, spin and parity $J^P = \frac{1}{2}^-$ [17] and the latest developments in amplitude analysis [18, 19], lattice QCD [20, 21, 22], and quark-diquark models [23, 24].

3.1. Status of amplitude analysis in the $\Lambda(1405)$ region

A recent combined amplitude analysis on $\bar{K}N$ scattering and $\pi\Sigma K^+$ photoproduction off the proton data based on the unitarization of the chiral potential via a coupled channel Bethe–Salpeter equation [18] has confirmed previous works [19, 25] that stated that the $\Lambda(1405)$ compiled in the RPP [1] is not a single state but two different resonances (poles) located at $1429^{+8}_{-7} - i\,12^{+2}_{-3}$ MeV and at $1325^{+15}_{-15} - i\,90^{+12}_{-18}$ MeV [18]. In [18] and previous analyses [25], the resonances are dynamically generated and their authors interpreted both poles as molecular states whose origin is entirely due to the $\bar{K}N$ and $\pi\Sigma$ interactions. However, an interpretation of this kind has to be taken with care because, besides model-dependency, the employed chiral Lagrangian in [18] incorporates fourteen low-energy constants that were treated as free parameters with the only requirement of being of natural size. These constants encapsulate the information on unresolved degrees of freedom and can hide resonant states whose origin is not dynamical.

3.2. Status of $\Lambda(1405)$ *in quark models*

The $\Lambda(1405)$ quantum numbers are those of a *uds* state but, in the past, constituent quark models have failed to match the experimental mass [26]. This situation has improved during 2015 with two calculations based on the relativistic interacting quark-diquark model that obtain values closer to the amplitude extractions: 1431 MeV in [23] and 1406 MeV in [24].

3.3. Status of $\Lambda(1405)$ *in lattice QCD*

Although it would be desirable to understand the nature of the $\Lambda(1405)$ directly from QCD via lattice QCD, the resonant nature of this state makes its numerical investigation taxing. There have been recent exploratory studies of this state from lattice QCD identifying the $\Lambda(1405)$ as a $\bar{K}N$ molecule in [21] and as an ordinary three-quark state in [22]. Unfortunately, in accessing the spectrum and the subsequent analysis of the spectrum the strong decay to the $\pi\Sigma, \bar{K}N$ scattering states is neglected. This approximation has been formally [27] and numerically [28, 29] demonstrated to only be justifiable in the narrow-width limit. In fact, it has been recently demonstrated that finite volume matrix elements of electromagnetic current obtained via lattice QCD, cannot be directly interpreted as infinite-volume form factors of resonances [30] Hence, it is hard to asses the size of the systematic errors of the exploratory investigations listed above. In order to make a definitive statement of the implications of lattice QCD regarding the true nature of the $\Lambda(1405)$, a thorough analysis of the coupled $\pi\Sigma, \bar{K}N$ scattering amplitude will be needed, as has been done in [29] for the mesonic sector.

3.4. Regge analysis of the Λ^* *spectrum*

It is clear that a big progress has been made in the last years, but the final answer on the nature of the $\Lambda(1405)$ state(s) remains elusive. In [31] we used Regge phenomenology and our knowledge on the hyperon spectrum to assess the nature of the two $\Lambda(1405)$ poles. From complex angular momentum theory we know that resonances follow Regge trajectories that connect poles at different partial waves [32]. Hence, we can use Regge trajectories to connect the ground states to their excitations and derive conclusions based on the shape of the trajectories, as their (non-)linearity depend on the quark-gluon dynamics [33]. If we use the current knowledge on the Λ^* spectra we can compute the parent Regge trajectories and shed light on the nature of the $\Lambda(1405)$ states by testing how well do the two poles fit within said trajectories. In figure 3 we show the Chew–Frautschi plot [34], *i.e.* $\Re(s_p)$ *vs.* J, where s_p is the pole position and J stands for the total angular momentum of the resonance. The $\Lambda(1116)$ is taken from the RPP [1], $\Lambda(1405)_a$ ($1429^{+8}_{-7} - i\,12^{+2}_{-3}$ MeV) and $\Lambda(1405)_b$ ($1325^{+15}_{-15} - i\,90^{+12}_{-18}$ MeV) from [18], and the rest of the Λ^* poles from the $\bar{K}N$ amplitude analysis presented in section 2. It is apparent how the projection of the Regge trajectories onto the ($\Re(s_p)$,J) plane in figure 3 follow linear trajectories (with some small deviations). At first order approximation we expect a linear behavior for three-quark states in the energy region where the resonances we have studied lie [33]. However, both $\Lambda(1405)$ resonances lay on top of the parent Regge trajectory. The situation clarifies if we realize that the poles are complex quantities and it is also possible to plot $-\Im(s_p)$ *vs.* J, which corresponds to the projection of the Regge trajectory onto the ($-\Im(s_p)$,J) plane. We provide such plot in figure 4. It is apparent how the poles follow a square-root behavior and that the $\Lambda(1405)_a$ state fits in the unnatural parity parent Regge trajectory while the $\Lambda(1405)_b$ state does not. Hence, $\Lambda(1405)_a$ shares the same nature of the other states of the unnatural parity parent Regge trajectory, *i.e.* it is mostly a three-quark state, while $\Lambda(1405)_b$ does not belong to any linear Regge trajectory and, hence, cannot be a three-quark state, *i.e.* it is either a molecule or a pentaquark. We refer the reader to [31] for a quantitative analysis and details on the arguments. This result is consistent with quark-diquark models finding only one $\Lambda(1405)$ state [23, 24] and different results in the exploratory lattice QCD calculations [21, 22]. Further studies should assess if the nature of $\Lambda(1405)_b$ is that of a pentaquark or a molecular state.

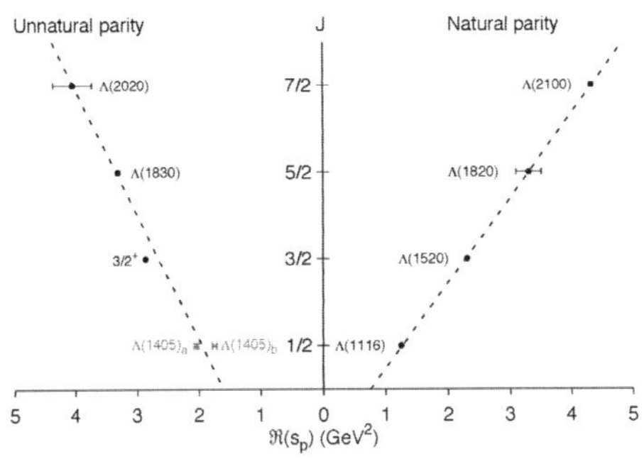

Figure 3. $\Re(s_p)$ *vs.* J (Chew–Frautschi) plot for the Λ parent Regge trajectories. The dashed lines are to guide the eye.

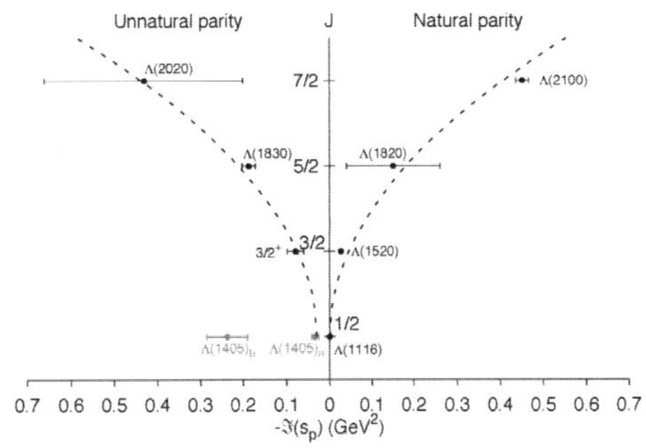

Figure 4. $-\Im(s_p)$ *vs.* J plot for the Λ parent Regge trajectories. The dashed lines are to guide the eye.

4. Conclusions

(i) We have developed a $\bar{K}N$ amplitude in the resonance region that incorporates up to 13 channels per partial wave, has the right angular momentum barrier, analyticity and unitarity, obtaining the most comprehensive picture of the $S = -1$ hyperon spectrum to the date;

(ii) The model can be incorporated in the analysis of three-body decay experiments for pentaquark searches and two kaon photoproduction experiments for strangeonia and exotics with hidden strangeness search;

(iii) The codes to compute the partial waves and the observables (cross sections and asymmetries) can be run online and downloaded from [15];

(iv) Using Regge phenomenology and our knowledge of the Λ^* spectrum we are able to establish that the higher-mass $\Lambda(1405)$ resonance is mostly a three-quark state while the lower-mass $\Lambda(1405)$ resonance is either a molecule or a pentaquark.

Acknowledgments

This work is part of the efforts of the Joint Physics Analysis Center. We thank Raúl Briceño for useful discussions. We thank the organizers for their invitation to the conference and their warm hospitality at Cocoyoc.

References

[1] Olive KA *et al* (Particle Data Group) 2014 *Chin. Phys.* C **38** 090001
[2] Qiang Y *et al* 2010 *Phys. Lett.* B **694** 123
[3] Fernández-Ramírez C *et al* 2016 *Phys. Rev.* D **93** 034029
[4] Zhang H, Tulpan J, Shrestha M and Manley DM 2013 *Phys. Rev.* C **88** 035205
Kamano H, Nakamura SX, Lee T-SH and Sato T 2014 *Phys. Rev.* C **90** 065204

Kamano H, Nakamura SX, Lee T-SH and Sato T 2015 *Phys. Rev.* C **92** 025205
[5] Aaij R *et al* (LHCb Collaboration) 2015 *Phys. Rev. Lett.* **115** 072001
[6] Briceño RA *et al* 2016 *Chin. Phys.* C **40** 042001
[7] Battaglieri M *et al* 2015 *Acta Phys. Polon.* B **46** 257
[8] Dudek JJ *et al* 2012 *Eur. Phys. J.* A **48** 187
[9] Wigner EP 1946 *Phys. Rev.* **70** 15
Badalyan AM, Kok LP and Simonov YA 1982 *Phys. Rep.* **82** 31
[10] Zhang H, Tulpan J, Shrestha M and Manley DM 2013 *Phys. Rev.* C **88**, 035204
[11] James F and Roos M 1975 *Comput. Phys. Commun.* **10** 343
[12] Fernández-Ramírez C, Moya de Guerra E, Udías A and Udías JM 2008 *Phys. Rev.* C **77** 065212
[13] Press WH, Teukolsky SA, Vetterling WT and Flannery BP 1992 *Numerical Recipes: The Art of Scientific Computing* (Cambridge: Cambridge University Press)
[14] Mathieu V 2016 *Preprint* 1601.01751 [hep-ph]
[15] `http://www.indiana.edu/~jpac/index.html`
[16] Dalitz RH and Tuan SF 1959 *Phys. Rev. Lett.* **2** 425
Alston MH *et al* 1961 *Phys. Rev. Lett.* **6** 698
Engler A *et al* 1965 *Phys. Rev. Lett.* **15** 224
[17] Moriya K *et al* 2014 *Phys. Rev. Lett.* **112** 082004
[18] Mai M and Meißner U-G 2015 *Eur. Phys. J.* A **51** 30
[19] Jido D *et al* 2003 *Nucl. Phys.* A **725** 181
Hyodo T and Weise W 2008 *Phys. Rev.* C **77** 035204
Roca L and Oset E 2013 *Phys. Rev.* C **87** 055201
[20] Melnitchouk W *et al* 2003 *Phys. Rev.* D **67** 114506
Nemoto Y, Nakajima N, Matsufuru H and Suganuma H 2003 *Phys. Rev.* D **68** 094505
Burch T *et al* 2006 *Phys. Rev.* D **74** 014504
Menaude BJ, Kamleh W, Leinweber DB and Mahbub MS 2012 *Phys. Rev. Lett.* **108** 112001
Molina R and Döring M 2015 *Preprint* 1512.05831 [hep-lat]
[21] Hall JMM *et al* 2015 *Phys. Rev. Lett.* **114** 132002
[22] Engel GP, Lang CB and Schäfer A 2013 *Phys. Rev.* D **87** 034502
Engel GP, Lang CB, Mohler D and Schäfer A 2013 *Phys. Rev.* D **87** 074504
[23] Santopinto E and Ferretti J 2015 *Phys. Rev.* C **92** 025202
[24] Faustov RN and Galkin VO 2015 *Phys. Rev.* D **92** 054005
[25] Dalitz RH and Tuan SF 1960 *Ann. Phys., NY* **10** 307
Dalitz RH, Wong TC and Rajasekaran G 1967 *Phys. Rev.* **153** 1617
Kaiser N, Waas T and Weise W 1997 *Nucl. Phys.* A **612** 297
Oller JA and Meißner U-G 2001 *Phys. Lett.* B **500** 263
Roca L, Hyodo T and Jido D 2008 *Nucl. Phys.* A **809** 65
Hyodo T and Jido D 2012 *Prog. Part. Nucl. Phys.* **67** 55
Roca L and Oset E 2013 *Phys. Rev.* C **88** 055206
[26] Capstick S and Isgur N 1986 *Phys. Rev.* D **34** 2809
Löring U, Metsch BC and Petry HR 2001 *Eur. Phys. J.* A **10** 447
Melde T *et al* 2008 *Phys. Rev.* D **77** 114002
[27] Lüscher M 1986 *Commun. Math. Phys.* **105** 153
Rummukainen K and Gottlieb SA 1995 *Nucl. Phys.* B **450** 397
Briceño RA, Hansen MT and Walker-Loud A 2015 *Phys. Rev.* D **91** 034501
Briceño RA and Hansen MT 2015 *Phys. Rev.* D **92** 074509
[28] Dudek JJ, Edwards RG and Thomas CE 2013 *Phys. Rev.* D **87** 034505
[29] Dudek JJ, Edwards RG, Thomas CE and Wilson DJ 2014 *Phys. Rev. Lett.* **113** 182001
Wilson DJ *et al* 2015 *Phys. Rev.* D **92** 094502
Dudek JJ, Edwards RG and Wilson DJ 2016 *Preprint* 1602.05122 [hep-ph]
[30] Briceño RA and Hansen MT 2015 *Preprint* 1509.08507 [hep-lat]
[31] Fernández-Ramírez C, Danilkin IV, Mathieu V and Szczepaniak AP 2016 *Phys. Rev.* D **93** 074015
[32] Gribov VN 2003 *The Theory of Complex Angular Momenta* (Cambridge: Cambridge University Press)
[33] Kahana D, Maung Maung K and Norbury JW 1993 *Phys. Rev.* D **48** 3408
Tang A and Norbury JW 2000 *Phys. Rev.* D **62** 016006
Inopin A and Sharov GS 2001 *Phys. Rev.* D **63** 054023
[34] Chew GF and Frautschi SC 1962 *Phys. Rev. Lett.* **8** 41

Tests of Predictions of the Algebraic Cluster Model: the Triangular $\mathcal{D}_{3h}$ Symmetry of ^{12}C

Moshe Gai
LNS at Avery Point, University of Connecticut, 1084 Shennecossett Rd., Groton, CT 06340
E-mail: moshe.gai@uconn.edu

Abstract. A new theoretical approach to clustering in the frame of the Algebraic Cluster Model (ACM) has been developed. It predicts rotation-vibration structure with rotational band of an oblate equilateral triangular symmetric spinning top with a $\mathcal{D}_{3h}$ symmetry characterized by the sequence of states: $0^+, 2^+, 3^-, 4^\pm, 5^-$ with a degenerate 4^+ and 4^- (parity doublet) states. Our measured new 2_2^+ in ^{12}C allows the first study of rotation-vibration structure in ^{12}C. The newly measured 5^- state and 4^- states fit very well the predicted ground state rotational band structure with the predicted sequence of states: $0^+, 2^+, 3^-, 4^\pm, 5^-$ with almost degenerate 4^+ and 4^- (parity doublet) states. Such a $\mathcal{D}_{3h}$ symmetry is characteristic of triatomic molecules, but it is observed in the ground state rotational band of ^{12}C for the first time in a nucleus. We discuss predictions of the ACM of other rotation-vibration bands in ^{12}C such as the (0^+) Hoyle band and the (1^-) bending mode with prediction of ("missing 3^- and 4^-") states that may shed new light on clustering in ^{12}C and light nuclei. In particular, the observation (or non observation) of the predicted ("missing") states in the Hoyle band will allow us to conclude the geometrical arrangement of the three alpha particles composing the Hoyle state at 7.6542 MeV in ^{12}C. We discuss proposed research programs at the Darmstadt S-DALINAC and at the newly constructed ELI-NP facility near Bucharest to test the predictions of the ACM in isotopes of carbon.

1. Introduction

A recent experiment performed at the HIγS gamma-ray facility [1] using the optical readout TPC (O-TPC) [2] shown in Fig. 1, provided unambiguous identification of the 2^+ member of the Hoyle rotational band [3]. This observation allows for the first time the study of the predicted rotation-vibration structure of ^{12}C.

The structure of light nuclei, and specifically of clustering received new interest with major developments of ab-initio calculations of light nuclei and in particular of ^{12}C. Theoretical ab-initio shell model [4] and symmetry inspired shell model [5] calculations as well as ab-intio Effective Field Theory (EFT) calculations on the lattice [6, 7], Fermionic Molecular Dynamics (FMD) model [8] and Anti-symmetrized Molecular Dynamics (AMD) model [9], are employed to yield a microscopic foundation of the clustering phenomena that naturally occurs in cluster models [10, 11]. For example, one issue of current concern [12] is the geometrical structure of the three alpha-particles in the Hoyle state at 7.6542 MeV in ^{12}C (linear chain, obtuse triangle or equilateral triangle) and the Hoyle rotational band built on top of the Hoyle state [13, 14, 15, 16, 17].

Cluster states are best described as molecular states which are characterized by the separation (Jacobi) vector(s) connecting the constituent objects. For diatomic like object one single

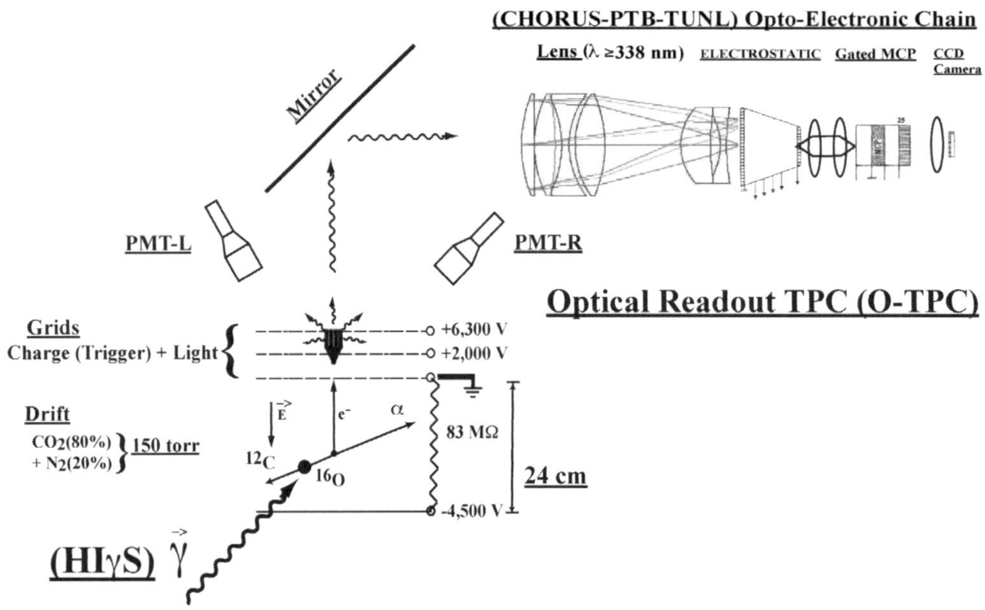

Figure 1. A schematic diagram of the Optical Readout Time Projection Chamber (O-TPC) [2]. The gas mixture and pressure as well as the operating voltages are indicated. The drift electron(s) that provide the third (time projection) upward dimension is shown schematically.

(separation) vector is required leading to the predicted U(4) symmetry [18] that was observed in ^{18}O via the characteristic enhanced E1 decays [19]. Triatomic symmetric spinning tops are characterized by the two perpendicular Jacobi vectors (λ and ρ) leading to the predicted U(7) symmetry with the geometrical $\mathcal{D}_{3h}$ symmetry [20, 21].

The observation of the 2^+ Hoyle rotational excitation in ^{12}C [3, 14] together with the recently discovered 4^- [22, 23] and 4^+ [24] states in ^{12}C are in agreement with the predicted spectrum of an oblate spinning top with a $\mathcal{D}_{3h}$ symmetry [20, 21]. It was predicted [20, 21] that the three alpha-particle system of ^{12}C leads to the ground state rotational band including the most unusual sequence of states: $0^+, 2^+, 3^-, 4^\pm, 5^-$. The new high spin 5^- state [25] as well the previously published 4^- state [22, 23] lead to a J(J+1) trajectory as predicted by this U(7) model [25] including the nearly degenerate 4^- and 4^+ states as shown in Fig. 2.

2. The Algebraic Cluster Model

The spectrum of ^{12}C predicted by the Algebraic Cluster Model (ACM) [20, 21] is shown in Fig. 3 where it is also compared to the measured spectrum of ^{12}C [25]. In addition to the ground state rotational band this U(7) model [20, 21] predicts the Hoyle state at 7.6542 MeV in ^{12}C to be the first vibrational breathing mode of the three alpha-particle equilateral configuration leading to the same rotational structure albeit with a larger moment of inertia (by a factor of 2). Recent measurements revealed the 2^+ [14] and 4^+ [24] members of the predicted Hoyle rotational band and we are currently searching [26, 27] for the 4^- predicted by the ACM to be nearly degenerated with the 4^+ state and the 3^- (broad) state that was suggested to lie between 11 and 14 MeV [22, 28]. The observation (or lack there) of these "missing" states will allow us to determine whether the Hoyle state is composed of three alpha-particles in an equilateral triangle arrangement [25, 8, 9] or an obtuse triangle [7] or whether it is better described as a

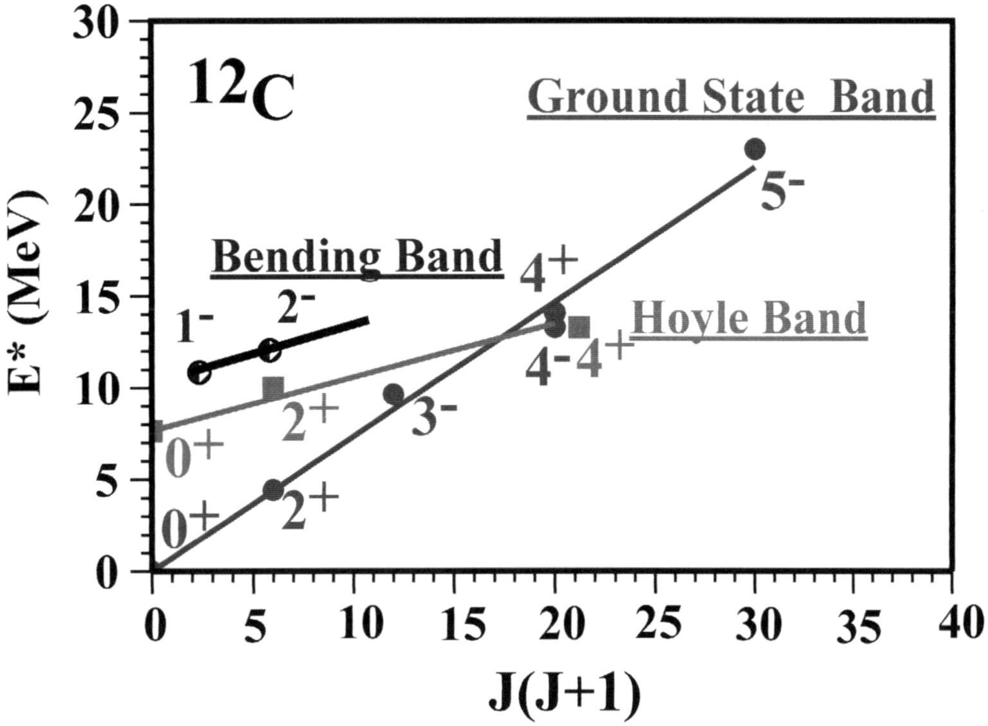

Figure 2. Rotational band structure of the: ground-state band, the Hoyle band and the bending vibration band in ^{12}C [25].

vibrational excitation of a "diffuse gas" of three alpha-particles [11].

The U(7) model also predicts the 1^- state at 10.84 MeV to be the vibrational bending mode with a rotational band including the 1^- and a degenerate $2^{\pm}$ states. We are searching [26, 27] for the third 2^+ of ^{12}C that is predicted by the U(7) model to lie near the observed 2^- state at 11.8 MeV.

3. Future Tests of the ACM

We propose a new research program in Nuclear Structure studies to be performed at the ELI-NP facility near Bucharest, Romania, with a newly proposed electronic readout TPC (eTPC) and a Silicon Strip Detector (SSD) [27]. We propose to measure of the multi-alpha decay of ^{12}C and ^{16}O with a TPC detector. Such measurements of multi-alpha decay of ^{12}C were also performed in the past with Silicon Strip Detectors (SSD) [22, 23, 24] hence similar measurements with SSD are also considered for the ELI-NP. We emphasize that the measurements we discuss in ^{12}C and ^{16}O are considered as typical examples of studies in nuclear structure that can be performed with our proposed detectors at the ELI-NP and such measurements in other light nuclei will lead to new perspective on the clustering phenomena in light nuclei.

4. Conclusions

In conclusion the ACM appears to open a new chapter in cluster physics of light nuclei and it presents an opportunity for further experimental investigation of light nuclei. We refer the

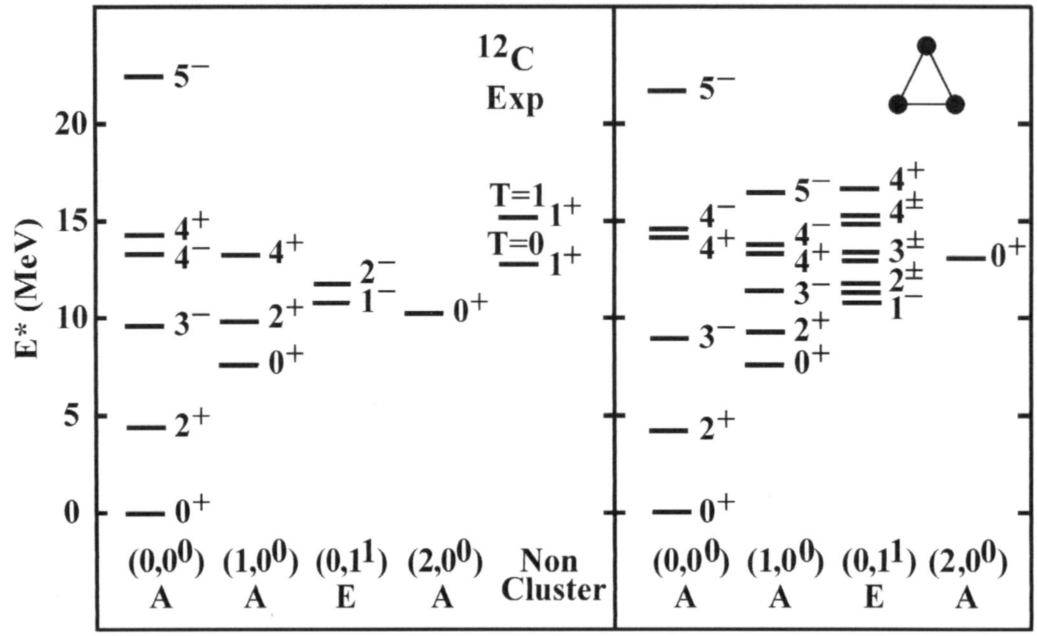

Figure 3. Comparison between the low-lying experimental spectrum of ^{12}C [25] and the energies of the oblate symmetric top. The levels are organized in columns corresponding to the ground state band and the vibrational bands with A and E symmetry [20, 21] of an oblate top with triangular symmetry. The last column on the left-hand side, shows the lowest observed non-cluster (1^+) levels.

reader to the Technical Design Report (TDR) [27] for a discussion of the scientific goals of the Charged Particle Working Group (CPWG) of the ELI-NP facility in the studies of light nuclei.

Acknowledgments

The material presented in this paper is based upon work supported by the U.S. Department of Energy, Office of Science, Office of Nuclear Physics, under Award Number DE-FG02-94ER40870.

References

[1] Weller H R, Ahmed M W, Gao H, Tornow W, Wu Y K, Gai M, and Miskimen R 2009 *Prog. Part. Nucl. Phys.* **62** 257
[2] Gai M, Ahmed M W, Stave S C, Zimmerman W R, Breskin A, Bromberger B, Chechik R, Dangendorf V, Delbar Th, France III R H, Henshaw S S, Kading T J, Martel P P, McDonald J E R, Seo P N, Tittelmeier K, Weller H R and Young A H 2010 *JINST* **5** 12004
[3] Zimmerman W R, Ahmed M W, Bromberger B, Stave S C, Breskin A, Dangendorf V, Delbar Th, Gai M, Henshaw S S, Mueller J M, Sun C, Tittelmeier K, Weller H R, and Wu Y K 2013 *Phys. Rev. Lett.* **110** 152502
[4] Roth R, Langhammer J, Calci A, Binder S, and Navratil P 2011 *Phys. Rev. Lett.* **107** 072501
[5] Dreyfuss A C, Launey K D, Dytrych T, Draayer J P, and Bahri C 2013 *Phys. Lett. B* **727** 511
[6] Epelbaum E, Krebs H, Lee D, and Meissner U G 2011 *Phys. Rev. Lett.* **106** 192501
[7] Epelbaum E, Krebs H, Lahde T A, Lee D, and Meissner U G 2012 *Phys. Rev. Lett.* **109** 252501
[8] Chernykh M, Feldmeier H, Neff T, Von Neumann-Cosel P, and Richter A 2007 *Phys. Rev. Lett.* **98** 032501
[9] Kanada-En'yo Y 2007 *Prog. Theor. Phys.* **117** 655
[10] Kamimura M 1981 *Nucl. Phys. A* **351** 456
[11] Funaki Y, Horiuchi H, Von Oertzen W, Ropke G, Schuck P, Tohsaki A, and Yamada T 2009 *Phys. Rev. C* **80** 064326

[12] Fynbo H O U and Freer M 2011 *Physics* **4** 94
[13] Morinaga H 1956 *Phys. Rev.* **101** 254
[14] Itoh M *et al.* 2004 *Nucl. Phys. A* **738** 268
Itoh M *et al.* 2011 *Phys. Rev. C* **84** 054308
[15] Freer M *et al.* 2009 *Phys. Rev. C* **80** 041303(R)
[16] Zimmerman W R, Destefano N E, Freer M, Gai M, and Smit F D 2011 *Phys. Rev. C* **84** 027304
[17] Freer M *et al.* 2012 *Phys. Rev. C* **86** 034320
[18] Iachello F 1981 *Phys. Rev. C* **23** 2778
[19] Gai M, Ruscev M, Hayes A C, Ennis J F, Keddy R, Schloemer E C, Sterbenz S M, and Bromley D A 1983 *Phys. Rev. Lett.* **50** 239
[20] Bijker R and Iachello F 2000 *Phys. Rev. C* **61** 067305
[21] Bijker R and Iachello F 2002 *Ann. Phys. (N.Y.)* **298** 334
[22] Freer M *et al.* 2007 *Phys. Rev. C* **76** 034320
[23] Kirsebom O S *et al.* 2010 *Phys. Rev. C* **81** 064313
[24] Freer M *et al.* 2011 *Phys. Rev. C* **83** 034314
[25] Marín-Lámbarri D J, Bijker R, Freer M, Gai M, Kokalova Tz, Parker D J, and Wheldon C 2014 *Phys. Rev. Lett.* **113** 012502
[26] Von Neumann-Cosel P 2013 *A 12C(e,e') experiment in progress at Darmstadt* Private communication
[27] Filipescu D, Anzalone A, Balabanski D L, Belyshev S S, Camera F, La Cognata M, Constantin P, Csige L, Cuong P V, Cwiok M, Derya V, Dominik W, Gai M, Gales S, Gheorghe I, Ishkhanov B S, Krasznahorkay A, Kuznetsov A A, Mazzocchi C, Orlin V N, Pietralla N, Sin M, Spitaleri C, Stopani K A, Tesileanu O, Ur C A, Ursu I, Utsunomiya H, Varlamov V V, Weller H R, Zamfir N V, and Zilges A 2015 *Eur. Phys. J. A* **51** 185
[28] Kirsebom O S 2015 *CHANDA, workshop on target preparations* Paul Scherrer Institute, Viligen, Switzerland

Resonance and non-resonance effect of continuum states of ^{6}Li on elastic scattering angular distributions

A. Gómez Camacho

Departamento de Aceleradores, Instituto Nacional de Investigaciones Nucleares, Apartado Postal 18-1027, C.P. 11801, México, D.F., Mexico

Abstract. CDCC calculations of elastic scattering angular distributions for reactions of the weakly bound projectile ^{6}Li with targets ^{28}Si and ^{58}Ni at energies around the Coulomb barrier are presented. Special emphasis is given to account for the effect of couplings from ^{6}Li resonance states $l = 2, J^\pi = 3^+, 2^+, 1^+$. Similarly, the effect produced by non-resonant state couplings is studied. The convergent calculations are carried out with global α-target and d-target interactions. The calculated elastic scattering angular distributions are in general in good agreement with the measurements for the systems considered in this work. It is found that the calculations with only resonance states are very similar to that with all couplings (resonance+non-resonance). So, the absence of these states have a strong effect on elastic scattering (non-resonance states calculation). It is shown that the effects increase as the collision energy increases. An interpretation of the strength of the different effects is given in terms of the polarization potentials that emerge from the different couplings.

1. Introduction

Reaction mechanisms involving weakly bound nuclei, both stable and radioactive, have been a subject of strong research [1, 2, 3]. One of the most interesting topics, is the effect of breakup on elastic scattering and fusion reaction mechanisms. It is now known that couplings of the elastic channel to continuum breakup states of the weakly bound projectile, as well as, continuum-continuum couplings are essential to fit the experimental data. This is so since, couplings to and between continuum states give rise to a strong repulsive polarization potential that decreases fusion absorption. Besides the non-capture breakup process, other nuclear mechanisms are present for reactions with weakly bound nuclei. For instance, sequential and direct complete fusion, incomplete fusion and nucleon transfer. Coupled channel calculations show that couplings between the incident elastic channel to fusion ones, give place to an attractive dynamical polarization potential V_F that lowers the nominal Coulomb barrier V_B. So, the net effect of V_F is to enhance fusion mainly for energies around the barrier. Similarly, couplings to breakup channels produce a repulsive dynamical polarization potential V_{BU} , that is particularly strong at energies below the barrier. In this energy range, V_{BU} overcomes V_F, so that the net effect is twofold: a) An increase in the fusion barrier, that leads to a net fusion suppression, and b) Most of the reaction cross section is accounted for the breakup process. Probably, one of the most appropriate method to calculate the effect of breakup on elastic and fusion reaction mechanisms of weakly bound systems is the Continuum Discretized Coupled-Channel (CDCC)

[4, 5]. This approach has been applied to a large range of weakly bound nuclear systems, for instance; $2n$-halo ^{6}He on ^{59}Co and ^{208}Pb [6, 7], ^{6}Li with targets ^{28}Si, ^{59}Co, ^{58}Ni, ^{144}Sm and ^{208}Pb [4, 6, 8, 9, 10, 11, 12, 13]. Also, for ^{7}Li with ^{28}Si [14, 15] and ^{144}Sm [16], the n-halo ^{11}Be with ^{208}Pb [17] and the p-halo ^{8}B with ^{58}Ni [10, 18]. In CDCC calculations for reactions with the weakly bound ^{6}Li, this is portrayed with a $\alpha - d$ cluster structure [6, 19]. Ground, resonant ($l = 2$, $j^\pi = 3^+, 2^+, 1^+$) and non-resonant states are included. Resonance states are identified with energies, $\varepsilon_{3+} = 0.71$ MeV, $\varepsilon_{2+} = 2.83$ MeV, $\varepsilon_{1+} = 4.18$ MeV (above threshold) and widths $\Gamma_{3+} = 0.024$ MeV, $\Gamma_{2+} = 1.7$ MeV and $\Gamma_{1+} = 1.5$ MeV respectively.

In the CDCC model, the continuum space is discretized in a finite number of states up to a maximum energy ϵ_{max}, the energy discretization is carried out with energy step $\Delta\epsilon$, maximum relative angular momentum l_{max} between the breakup fragments where tests of convergence of the calculations respect to these quantities are performed. The purpose of the present work is to calculate the effect on elastic scattering angular distributions, of the resonant continuum states of the weakly bound nucleus ^{6}Li$\rightarrow d + \alpha$, in reaction with the ^{28}Si and ^{58}Ni targets at energies around the Coulomb barrier. This is achieved by following two different processes; (a) omitting the discrete states belonging to the resonances from the whole continuum spectrum, b) Consider only resonance discrete states, that is omit non-resonance ones. In this manner, we can disentangle the importance of couplings to resonance states in the CDCC calculations. On the same footing, the polarization potentials associated with couplings to resonance states are calculated and the effect of resonance couplings on elastic scattering is explained in terms of these potentials.

Systematic global d-target and α-target interactions, which depend on the target mass A and incident collision energy E_{lab} are used in the calculations. Woods-Saxon potentials of Ref. [20] are used for the d-target interaction V_{dT}. The parameterization of these potentials has been obtained by a large analysis of experimental elastic and inelastic cross sections for reactions of deuteron with a large variety of nuclear target masses ($12 < A < 238$) and of incident energies ($E_{lab} < 183$ MeV). On the other hand, the density dependent double-folding Sao Paulo potential (SPP) [21] with adequate α-particle mass densities, is used to describe the α-target interaction $V_{\alpha T}$. The non-local SPP has been widely used with success to describe elastic and inelastic scattering, as well as, nucleon transfer cross sections for a number of nuclear systems. An extensive systematics of nuclear densities has been performed in Ref. [22] to produce a parameter-free interaction. In our calculations, the α-target interaction depends also only on the target mass A and kinetic energy of the α-particle respect to the initial incident energy of ^{6}Li. As for the $\alpha - d$ cluster structure of ^{6}Li, use is made of the Woods-Saxon parameterization given in Ref. [19]. Ground, unbound resonant and non-resonant continuum scattering states of ^{6}Li are generated with this interaction.

The paper is organized as follows. In section 2, a brief description of the CDCC formalism is given. Section 3 addresses the CDCC calculations for elastic scattering angular distributions, polarization potentials and how the effect of resonant continuum states of ^{6}Li is calculated. Finally, a summary is given in section 4.

2. CDCC formalism for a three-body system

A detailed description of the CDCC formalism is given in Refs.[4, 5]. In this paper we present only the basic equations that are required to perform our calculations. We consider the two-body cluster model structure of ^{6}Li (α-d) with ground state energy $E_{g.s.} = -1.47$ MeV. The model space for continuum states is that given in Ref. [19], where convergence at energies above the barrier has been checked. The wave function for breakup continuum states of ^{6}Li reads as,

$$\psi^P_{lj}(\mathbf{r}, k) = \{Y_{lm_l}(\hat{\mathbf{r}}) \otimes \chi_{I\mu s\sigma}\}_{lj} \frac{\varphi_{lsj}(r, k)}{r}, \tag{1}$$

where the internal wave function of the α-d system is $\chi_{I\mu s\sigma}$, with $I = 0$ and $s = 1$. $\varphi_{lsj}(r,k)$ describes the α-d relative radial motion with asymptotic wave-number k, orbital angular momentum l and total angular momentum j.

The continuum states $\varphi_{lsj}(r,k)$ in Eq. (1) are not square-integrable. However, Ref. [4] provides a recipe for constructing square-integrable wave functions known as bin states. A bin state, $u_{\beta}^{(i)}(r)$, is made of scattering wave functions within a given interval i, of continuum k values, $k_{i-1} < k < k_i$, *i.e.*,

$$u_{\beta=lsj}^{(i)}(r) = \sqrt{\frac{2}{\pi\eta}} \int_{k_{i-1}}^{k_i} w_i(k,l) e^{-i\delta_k(l)} \varphi_{lsj}(r,k) dk, \tag{2}$$

where $\delta_k(l)$ are scattering phase-shifts of φ_β and $w_i(k,l)$ are weight functions defined by,

$$\eta_l = \int_{k_{i-1}}^{k_i} |w_i(k,l)|^2 \, dk. \tag{3}$$

Actually, the weight functions w_i associated to non-resonant bin states are usually set as $w_i(k,l) = 1$, while for resonant states $w_i(k,l) = sin[\delta_k(l)]$.

The total wave function of the three-body system (α-d-$target$) is given by,

$$\Psi(\mathbf{R},\mathbf{r}) = \sum_{\beta} F_\beta(\mathbf{R}) \psi_\beta^P(\mathbf{r}) \Phi_0^T, \tag{4}$$

where Φ_0^T and $F_\beta(\mathbf{R})$ correspond to the inert target and projectile-target relative motion in the β-channel respectively. Projecting the equation of motion of the system onto the projectile states β, the following coupled equations are obtained,

$$\left[\hat{T}_R + U_{\beta\beta}^{(J)}(R) - (E - \varepsilon_\beta - \varepsilon_T)\right] F_\beta^{(J)}(R) = -\sum_{\beta'} U_{\beta\beta'}^{(J)}(R) F_{\beta'}^{(J)}(R). \tag{5}$$

Here, ε_T and ε_β are the internal energies of the target and projectile in the β-state. $U_{\beta\beta}^{(J)}(R)$ and $U_{\beta\beta'}^{(J)}(R)$ are the radial dependent diagonal and non-diagonal matrices of the interaction potentials $\hat{V}_{dT} + \hat{V}_{\alpha T}$. These are given by,

$$U_{\beta\beta'}^{(J)}(R) = < u_\beta \left| \hat{V}_{dT} + \hat{V}_{\alpha T} \right| u_{\beta'} >, \tag{6}$$

where the integration is carried out over the internal radial coordinate $\mathbf{r}$ and the angular coordinates of $\mathbf{R}$. The left-hand side of Eq.(6) are diagonal continuum couplings, in particular the term $U_{00}^{(J)}(R)$ corresponds to the optical potential in the elastic channel. The right-hand side of Eq.(6) corresponds to off-diagonal continuum couplings.

3. Effect of non-resonance and resonance states

In this section CDCC calculations of elastic scattering angular distributions are presented for the projectile ^{6}Li with targets ^{28}Si and ^{58}Ni, for incident energies just above the corresponding Coulomb barriers. The calculations are performed with the code FRESCO [23], where care is taken of the radial parameters R_j of the V_{dT} interaction. This is so, since in the code FRESCO, the general relation $R_j = r_j(A_1^{1/3} + A_2^{1/3})$ is considered while a shorter $R_j = r_j A^{1/3}$ is used in Ref. [20] where $A_1 = 2$ and A the target mass. To calculate the ground, resonant ($l = 2$, $j^\pi = 3^+, 2^+, 1^+$) and non-resonant discrete breakup states of ^{6}Li, we used the parameterization given in Ref. [19] for the interaction $V_{\alpha d}(r)$.

Table 1. Calculated values (MeV) for resonance energies and widths of ^{6}Li

Res	E_{cal}	Γ_{cal}	$E_{\rm exp}$	$\Gamma_{\rm exp}$
3^+	0.73	0.034	0.716	0.024
2^+	3.09	1.3	2.84	1.7
1^+	4.67	4.2	4.18	1.5

The discretization of the continuum is made as follows; the maximum angular momentum for the relative motion of the $\alpha - d$ fragments is $l_{max} = 3$, larger values do not have any effect on the calculations. So, bin states for $l = 0$, $j^\pi = 1^+$ and $l = 1$, $j^\pi = 0^-, 1^-, 2^-$ are constructed with step $\Delta\varepsilon = 0.5$ MeV up to $\varepsilon_{\rm max} = 6.8$ MeV. Finer and variable steps are used for resonant states $l = 2$, $j^\pi = 3^+, 2^+, 1^+$, so as to obtain centroid excitation energies and widths close to the corresponding measured values. Table 1 shows the calculated values with the interaction $V_{\alpha d}$ of Ref. [19].

For bin states $l = 3$, $j^\pi = 4^+, 3^+, 2^+$, a larger step $\Delta\varepsilon = 1.0$ MeV is used. Convergence tests at $\varepsilon_{\rm max} = 7.0, 7.5$ and 8.0 MeV were done with no effect on elastic angular distributions. Similarly, larger steps $\Delta\varepsilon = 0.75$ and 1.0 MeV were used with no appreciable effect on elastic scattering. Coulomb potential multipoles are included up to $L_Q = 4$. As regards the normalization coefficients of the interaction $V_{\alpha T}$, we set these as $N_R = 1$ and $N_I = 0.78$. This

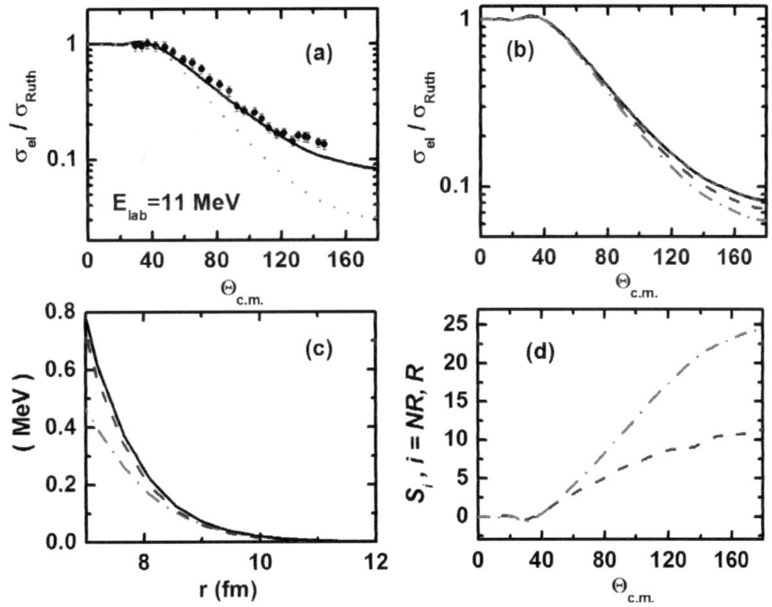

Figure 1. (Color online) (a) Elastic scattering angular distribution for ^{6}Li+^{28}Si (solid-line) at $E_{lab} = 11$ MeV and elastic channel calculation without couplings to continuum states (dotted line). Data taken from Refs. [24, 25, 26, 27]. (b) Calculations with non-resonant (dashed-dotted line) and resonant (dashed-line) discretized states. (c) Polarization potentials for the full (solid-line), non-resonant (dashed-dotted line) and resonant (dashed-line) discretized spaces. (d) Percentage effects of non-resonant (dashed-dotted line) and resonant (dashed-line) discretized states.

values are the result of a large systematic analysis of elastic scattering data of many nuclear systems performed by L.C. Chamon et al.,Refs. [21, 22]. So, here we assume these values.

The solid line in Fig.1a shows the CDCC calculation including all couplings, for elastic scattering of ^{6}Li with ^{28}Si at $E_{lab} = 11$ MeV. This energy $E_{c.m} = 9.05$ MeV is slightly above the Coulomb barrier $V_B = 6.7$ MeV. The experimental data are those reported in Refs. [24, 25, 26, 27]. The dotted line represents the calculation with only the elastic channel coupling, that is, only the term $U^{(J)}_{\beta=0\beta=0}(R)$ is considered in Eq.(6). This result, far from the data, was already expected, since it is well known that couplings to continuum states are essential to achieve agreement with the data. The dashed-dotted line of Fig.1b, represents the calculation when the excited states in the region of the resonances are ignored, that is, when couplings to and among these states are omitted in Eq.(5). It is observed that the omission of these states has some effect on elastic scattering (solid-line) at large angles. The dashed-line corresponds to the calculation where only couplings among resonant discretized are considered. This calculation is closer to the full calculation with all couplings. This result shows clearly that resonant state couplings play a very important role in the description of the elastic scattering data.

In Fig.1c, we show the results for the polarization potentials that emerge from couplings from the three kinds of calculations, i.e., all couplings (full-line), resonance states (dashed-line) and among non-resonance states (dashed-dotted line). The results were obtained using the prescription given in Ref. [23]. All polarization potentials have very small strength but the potential that emerges from calculations including only resonant states (dashed-dotted line) is more repulsive than that for non-resonant states (dashed-line).

Fig.1d shows the calculation for $S_i(\theta)$, defined as,

$$S_i(\theta) = \{[(\frac{\sigma_{el}}{\sigma_{Ruth}})_F(\theta) - (\frac{\sigma_{el}}{\sigma_{Ruth}})_i(\theta)]/(\frac{\sigma_{el}}{\sigma_{Ruth}})_F(\theta)\} \cdot 100, \quad (7)$$

where F represents the calculation with the full discretized space and $i = NR, R$, correspond to those calculations without the resonance discretized space and with only the resonance space, respectively. That is, $S_i(\theta)$ corresponds to the percentage disagreement from the full calculations for the elastic scattering cross sections to calculations without resonant states $S_{NR}(\theta)$ and with only resonant states $S_R(\theta)$, respectively. Clearly, the absence of resonance state couplings has a large effect on elastic scattering, being more important at large angles, when $S_{NR}(\theta)$ reaches a value of about 25%. However, when only resonant states are considered, the disagreement is smaller of about 10%.

Next, we investigate the energy dependence of resonance states by performing calculations at higher energies. Figs.2 shows the same sort of calculations as described above at $E_{lab} = 13$ MeV ($E_{c.m.} = 10.7$ MeV). The conclusions from Figs. 2a, 2b, 2c and 2d are very similar to those for the lowest energy. Fig.2a shows that a close result to the data is achieved and that couplings to continuum states are essential. In Fig.2d, it is observed that the maximum disagreement for S_{NR} is only slightly above 25% at backward angles, while S_R reaches a maximum of about 15%. Here, also the elastic scattering calculation with only resonant states represented by the dashed-line of Fig.2b is closer to the full calculation (solid-line) than the one without resonant states (dashed-dotted line). This is due to the more repulsive polarization potential for resonant states, as represented by the dashed-line of Fig.2c respect to the case without these states (dashed.dotted line). Although, the results up to here presented are for energies $E_{lab} = 11$ MeV and 13 MeV, which are above the barrier and are close to each other, it is easy to observe the increasing effect of the resonance states of ^{6}Li. As a matter of fact, at $E_{lab} = 13$ MeV, some oscillation of S_{NR} (green-dashed-dotted line) appears at large dispersion angles. These oscillations are even more evident at the higher energy $E_{lab} = 27$ MeV. Fig.3a shows that the CDCC calculations give a reasonable fit for dispersion angles less than 110 degrees. For larger angles the data present a strong diffraction pattern. However, the result with only resonance

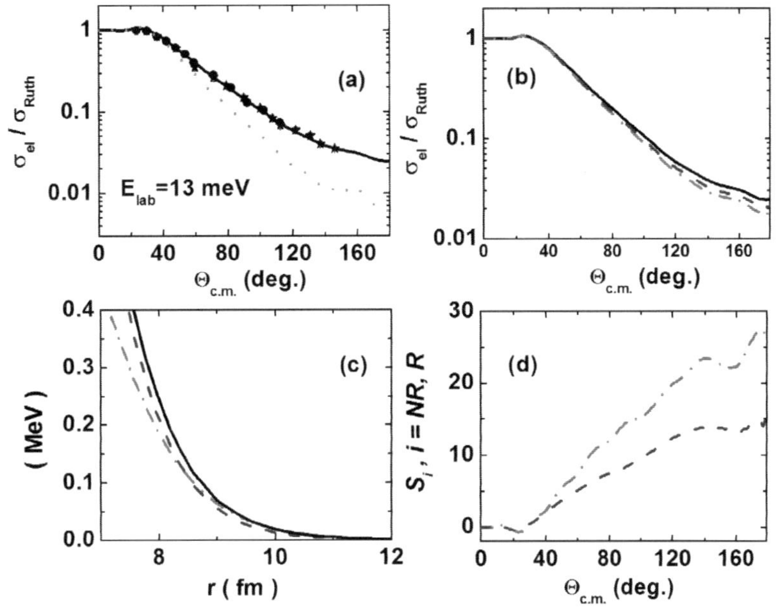

Figure 2. (Color online) Same as Fig.1 for ^{6}Li+^{28}Si at E_{lab} = 13 MeV.

states remains very close to the full calculations as seen by the dashed-line of Fig.3b, indeed, there is a maximum disagreement at about 80 degrees (19%) as the dashed-line of Fig.3d shows. On the other side, the non-resonance state calculation presents strong effects of nuclear (attractive) and Coulomb (repulsive) polarizations as shown by the dashed-dotted line of Fig.3d.

Figs.4 and 5 give the results for the energies E_{lab} = 12 and 18 MeV ($E_{c.m.}$ = 10.9 and 16.3 MeV) for the system ^{6}Li+^{58}Ni. These energies are below and above the barrier energy V_B = 12.4 MeV. The solid-lines of Figs.4a and 5a show a good agreement with data of Refs. [28, 29] and that couplings to continuum breakup states of ^{6}Li are very important to achieve these fits (dotted-lines). Also, the dashed-lines of Figs.4b and 5b indicate that resonance state couplings play a more important role than non-resonance ones for elastic scattering. This fact is explained by the more repulsive net polarization potentials corresponding to resonance states couplings alone as the dashed-lines of Figs.4c and 5c show. Finally, it is very interesting to observe that the effect of omitting resonance discrete states in the CDCC calculations of elastic scattering angular distributions increases as the collision energy increases. This is seen by the dashed-dotted lines of Figs.4d and 5d, where the percentage disagreement S_{NR} rises from around 3.5% to 30%. Similarly, it can be observed that the percentage disagreement S_R (dashed-lines), that is when only resonance states couplings are considered, also increases with the energy but at a lower rate. In this case, S_R rises from around 1% to 15% for the energies 12 and 18 MeV respectively. As a matter of fact, the shapes of S_{NR} and S_{NR} tend to be similar and to produce similar effects for higher energies. It is interesting to point out that the calculations of S_{NR} and S_R at 18 MeV show incipient oscillations at large dispersion angles, similar to those for ^{6}Li+^{28}Si at 27 MeV. Probably, it would be interesting to study this behavior for even higher energies.

The effects of resonance discretized states for the systems studied in this work ^{6}Li+^{28}Si and ^{6}Li+^{58}Ni are rather similar. For both, the maximum values of the functions S_R and S_{NR} are around 10$ and 30$ respectively for the energies studied. However, for the heavier target ^{144}Sm [30] at the energy E_{lab} = 30 MeV (V_B = 23.84 MeV), S_R and S_{NR} have values around 20% and

Journal of Physics: Conference Series **730** (2016) 012012 doi:10.1088/1742-6596/730/1/012012

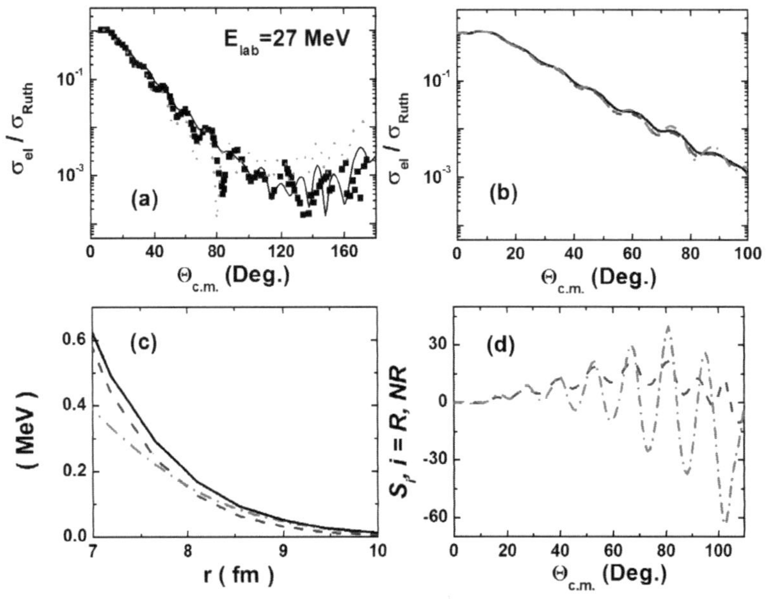

Figure 3. (Color online) Same as Fig.1 for ^{6}Li+^{28}Si at $E_{lab} = 27$ MeV.

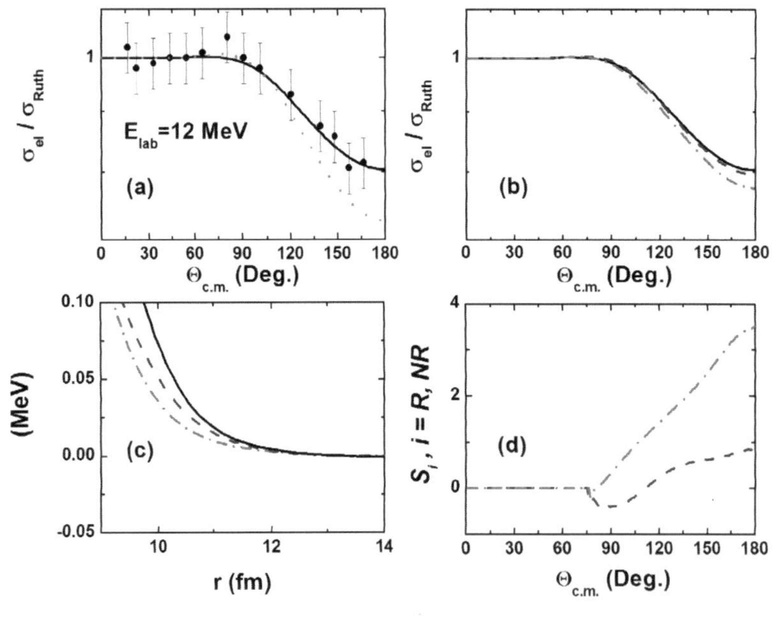

Figure 4. (Color online) Same as Fig.1 for ^{6}Li+^{58}Ni at $E_{lab} = 12$ MeV. Data from Refs. [28, 29].

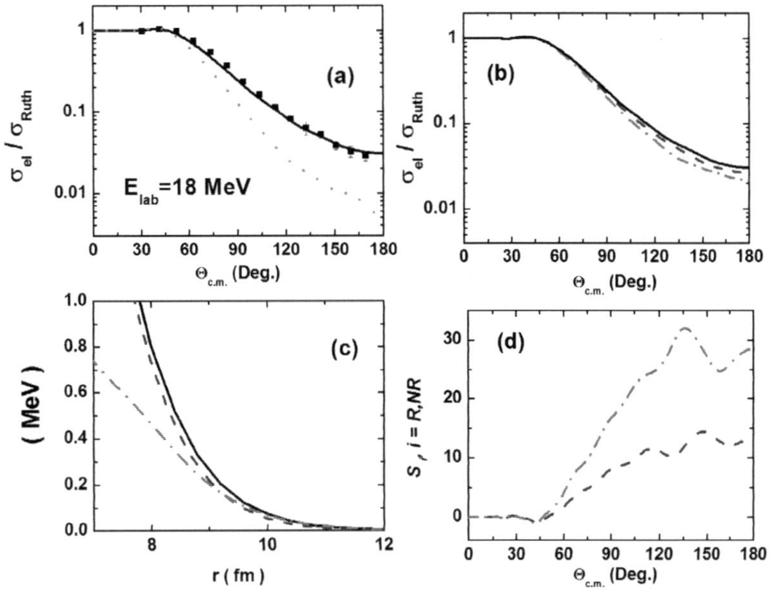

Figure 5. (Color online) Same as Fig.1 for ^{6}Li+^{58}Ni at $E_{lab} = 18$ MeV.

40% respectively. So, it is clear that the effects of the resonances of ^{6}Li depend on the target mass and are more evident for heavier targets than those considered in this work.

4. Summary and conclusions

CDCC calculations of elastic scattering angular distributions have been performed for the nuclear systems ^{6}Li+^{28}Si and ^{6}Li+^{58}Ni at energies around the Coulomb barrier. In the calculations, the cluster structure of the projectile ^{6}Li$\rightarrow \alpha + d$ is assumed, with global interactions for the α-target and d-target sub-systems. In general, good agreement between the calculations and the data has been achieved. For all of the energies studied in this work, it has been shown that calculations with to only the elastic channel are insufficient to fit the data, so, couplings to breakup states of ^{6}Li are very important. On the same footing, the effect on elastic scattering due to couplings to resonance states of ^{6}Li, namely, $l = 2$, $J^\pi = 3^+, 2^+$ and 1^+ has been calculated. This effect has been calculated by following two approaches, (a) by omitting from the whole energy discrete space the states corresponding to the resonances and (b) by considering only resonance state couplings. For both nuclear systems, the calculations for the case (a) show a strong effect so that in most cases the data are not fitted. However, the calculations with only resonance state couplings are closer to the data, showing that these states play a more important role for elastic scattering. The different effects of resonance and non-resonant state couplings on elastic scattering are better understood in terms of the polarization potentials that emerge from these couplings. It has been shown that resonance states produce stronger repulsive polarization potentials than non-resonance ones. In order to quantify the effect of resonance and non-resonance states, the percentage quantities, S_R and S_{NR} as functions of the dispersion angle $\theta_{c.m.}$ were studied. For the energies considered in this work, S_{NR} is mostly positive and reaches higher values as the energy increases. On the other hand, S_R shows also increasing values but at a lower rate. So, in the CDCC calculation of elastic scattering angular distributions, couplings from resonant states are more important than non-resonant ones at low collision energies. However, as the energy increases, resonant state couplings seem to result less significant and both calculations tend to produce similar effects.

References

[1] Canto L F, Gomes P R S, Donangelo R, and Hussein M S 2006 *Phys. Rep.* **424** 1
[2] Liang J F and Signorini C 2005 *Int. J. Mod. Phys. E* **14** 1121
[3] Keeley N, Raabe R, Alamanos N, and Sida J L 2007 *Prog. Part. Nucl. Sci.* **59** 579
[4] Sakuragi Y, Yahiro M, and Kamimura M 1986 *Prog. Theor. Phys. Suppl.* **89** 1
[5] Austern N, Iseri Y, Kamimura M, Kawai M, Rawitscher G, and Yahiro M 1987 *Phys. Rep.* **154** 125
[6] Beck C, Keeley N, and Diaz-Torres A 2007 *Phys. Rev. C* **75** 054605
[7] Fernandez-Garcia J P *et al.* 2010 *Phys. Lett. B* **693** 310
[8] Zerva K *et al.* 2009 *Phys. Rev. C* **80** 017601
[9] Rusek K, Keeley N, Pakou A, and Alamanos A 2007 *Nucl. Phys. A* **784** 13
[10] Keeley N, Mackintosh R S, and Beck C 2010 *Nucl. Phys. A* **380** 1
[11] Keeley N and Rusek K 1996 *Phys. Lett. B* **375** 9
[12] Otomar D R, Lubian J, and Gomes P R S 2010 *Eur. Phys. J. A* **46** 285
[13] Keeley G R *et al.* 2000 *Phys. Rev. C* **63** 024601
[14] Pakou A *et al.* 2004 *Phys. Rev. C* **69** 054602
[15] Pakou A *et al.* 2009 *Eur. Phys. J A* **39** 187
[16] Otomar D R *et al.* 2009 *Phys. Rev. C* **80** 034614
[17] Canto L F, Lubian J, Gomes P R S, and Hussein M S 2009 *Phys. Rev. C* **80** 047601
[18] Lubian J, Correa T, Aguilera E F, Canto L F, Gomez Camacho A, Quiroz E M, and Gomes P R S 2009 *Phys. Rev. C* **79** 064605
[19] Diaz-Torres A, Thompson I J and Beck C 2013 *Phys. Rev. C* **68** 044607
[20] An H and Cai C 2006 *Phys. Rev. C* **73** 054605
[21] Chamon L C *et al.* 1997 *Phys. Rev. Lett.* **79** 5218
[22] Chamon L C *et al.* 2002 *Phys. Rev. C* **66** 014610
[23] Thompson I J 1988 *Comput. Phys. Rep.* **7** 167
[24] Pakou A *et al.* 2003 *Phys. Lett. B* **556** 21
[25] Poling J E, Norbeck E, and Carlson R R 1976 *Phys. Rev. C* **13** 648
[26] Cook J 1982 *Nucl. Phys. A* **375** 238
[27] Vineyard M F, Cook J, and Kemper K W 1983 *Nucl. Phys. A* **405** 429
[28] Aguilera E F *et al.* 2009 *Phys. Rev. C* **79** 021601(R)
[29] Pfeiffer K O, Speth E, and Bethge K 1973 *Nucl. Phys. A* **206** 545
[30] Gómez Camacho A, Diaz-Torres A, Gomes P R S, and Lubian J 2016 *Phys. Rev. C* **93** 024604

Journal of Physics: Conference Series **730** (2016) 012013 doi:10.1088/1742-6596/730/1/012013

Latest news from the High Altitude Water Cherenkov Observatory

A González Muñoz for the HAWC Collaboration

Instituto de Física, Universidad Nacional Autónoma de México, Apartado Postal 20-364, México D.F. 01000, México

E-mail: adiv.gonzalez@fisica.unam.mx

Abstract. The High Altitude Water Cherenkov Observatory is an air shower detector designed to study very-high-energy gamma rays ($\sim$ 100 GeV to $\sim$ 100 TeV). It is located in the Pico de Orizaba National Park, Mexico, at an elevation of 4100 m. HAWC started operations since August 2013 with 111 tanks and in April of 2015 the 300 tanks array was completed. HAWC's unique capabilities, with a field of view of $\sim$ 2 sr and a high duty cycle of 5%, allow it to survey 2/3 of the sky every day. These features makes HAWC an excellent instrument for searching new TeV sources and for the detection of transient events, like gamma-ray bursts. Moreover, HAWC provides almost continuous monitoring of already known sources with variable gamma-ray fluxes in most of the northern and part of the southern sky. These observations will bring new information about the acceleration processes that take place in astrophysical environments. In this contribution, some of the latest scientific results of the observatory will be presented.

1. Introduction

The direct detection of high energy gamma rays is only possible with instruments installed in satellites orbiting the Earth (like Fermi-LAT [1]). However, the effective area of these instruments is limited, mainly due to the high costs of delivering payloads into orbit, which makes quite challenging to detect gamma rays of higher energies. However, very-high energy (VHE) gamma rays (as well as cosmic rays) upon interaction with the Earth atmosphere will produce cascades of secondary particles called Extensive Air Showers (EAS). Most of these secondary particles will lose all of its energy crossing the atmosphere but some of them will reach the ground. The EAS, being composed by particles traveling at relativistic speeds, will produce Cherenkov light in the air. This Cherenkov light can be collected and concentrated by reflectors into sensitive cameras, from which and "image" of the shower can be extracted. From this image the energy and direction of the primary gamma ray can be estimated. Instruments that use this technique for the observations of VHE gamma rays are called Imaging Atmospheric Cherenkov Telescopes (IACTS). The other main technique to indirectly observe VHE gamma rays is by detecting the particles from the EAS that hit the ground. A relatively cheap way to do it is by measuring the Cherenkov light that these particles produce in water. The pioneer of this technique for the detection of VHE gamma rays was *Milagro* [2], which operated from 2000 to 2008. It consisted of a rectangular pond of 80 m long, 60 m wide and 8 m deep, located at 2630 m above sea level in the Jemez Mountains, New Mexico, U.S.A. Inside the pond there were 723 photomultiplier tubes (PMTs) arranged in a grid. By registering the time in which the particles hit the pond and the intensity of the Cherenkov light that they produced it was possible to estimate the direction and the energy of the primary gamma ray. In order to achieve better sensitivity with this technique, the detector has to be as close as possible to the maximum of the EAS and cover a big area. With these principles in mind the High Altitude Water Cherenkov (HAWC) observatory was constructed.

2. HAWC Observatory

The HAWC observatory was inaugurated on March 20th, 2015. It was designed to be sensitive to gamma rays and cosmic rays between the energies of 100 GeV and 100 TeV [3]. The facility consists of an array of 300 water Cherenkov detectors (WCDs) deployed in a surface of $\sim$ 22000 m^2 at 4100 m above sea level, in a valley between the volcanoes Pico de Orizaba and Sierra Negra, in Mexico. Each WCD consist of a cylindrical structure of steel with 7.3 m of diameter and 4.5 m high. Inside of the steel cylinder there is a light-tight bladder where 200 000 l of highly purified water are stored. Each WCD is instrumented with four photomultiplier tubes (PMTs) installed at the bottom of the bladder, facing upward. Three of these PMTs (8-in photocathode) are deployed forming an equilateral triangle (side length 3.2 m), centered around the forth one (10-in photocathode) located at the centre of the WCD.

The secondary particles in an EAS are distributed around the shower axis in what is called the "shower front". These particles hit the WCDs of HAWC at different times, depending on the orientation of the shower axis. Additionally, the Cherenkov light produced in the WCDs depends on the flux intensity of the shower. With this information (time and charge) the direction and energy of the primary gamma ray is possible to be reconstructed. HAWC has an instantaneous field of view of 2 sr, allowing it to scan 2/3 of the sky per day.

3. HAWC's first scientific results

One of the first and most interesting scientific results from the HAWC observatory, is the detection of a small-scale anisotropy in the arrival direction of TeV cosmic rays. The data used for this study was gathered between June 16, 2013 and February 27, 2014 [4]. In order to find the small-scale anisotropy, first, a reference map is created in which asymmetries due to diurnal changes coming from changes in the cosmic-ray rate and the asymmetric shape are taken into account. Then, a relative intensity map is created computing the amplitude of deviations from the isotropic expectation (see Eq. 2 in [4]). Then, to see structures of scales smaller than 60°, the dipole, quadrupole and octupole terms are removed from this map, creating the maps shown in Fig. 1. Three regions are clearly visible, labeled as region A, B and C. These regions are largely in agreement with other anisotropy measurements by Milagro [2], Tibet ASγ [5] and ARGO-YBJ [6], in the same regions scanned by HAWC. For a more detailed explanation of the methods used and discussion of the results please see [4].

Another important result from HAWC is the identification of sources and candidate sources in the inner galactic plane (longitude [+15°,+50°] and latitude [−4°,+4°] in galactic coordinates) using data from August 2, 2013 to July 9, 2014 [7]. A skymap was created in the region of interest containing all the excess events over a normalized background. Then, to identify the number of sources needed to model the excess events, a likelihood ratio test is performed, first with only 1 source, then with 2 and so on, in a iterative process. The model for the source spectrum is assumed to follow a simple power-law function, with a fixed index of 2.3. Fig. 2 shows the significance skymap and the model for 11 sources. Three sources have been identified with more than 5 σ of significance and another seven candidates with more than 3 σ. For a more detailed discussion please see [7].

4. HAWC monitoring of the gamma-ray sky

One of the most notorious capabilities of the HAWC observatory is its high duty cycle ($\sim$ 95 %). Together with its large field of view, makes it the best instrument in current operation for an

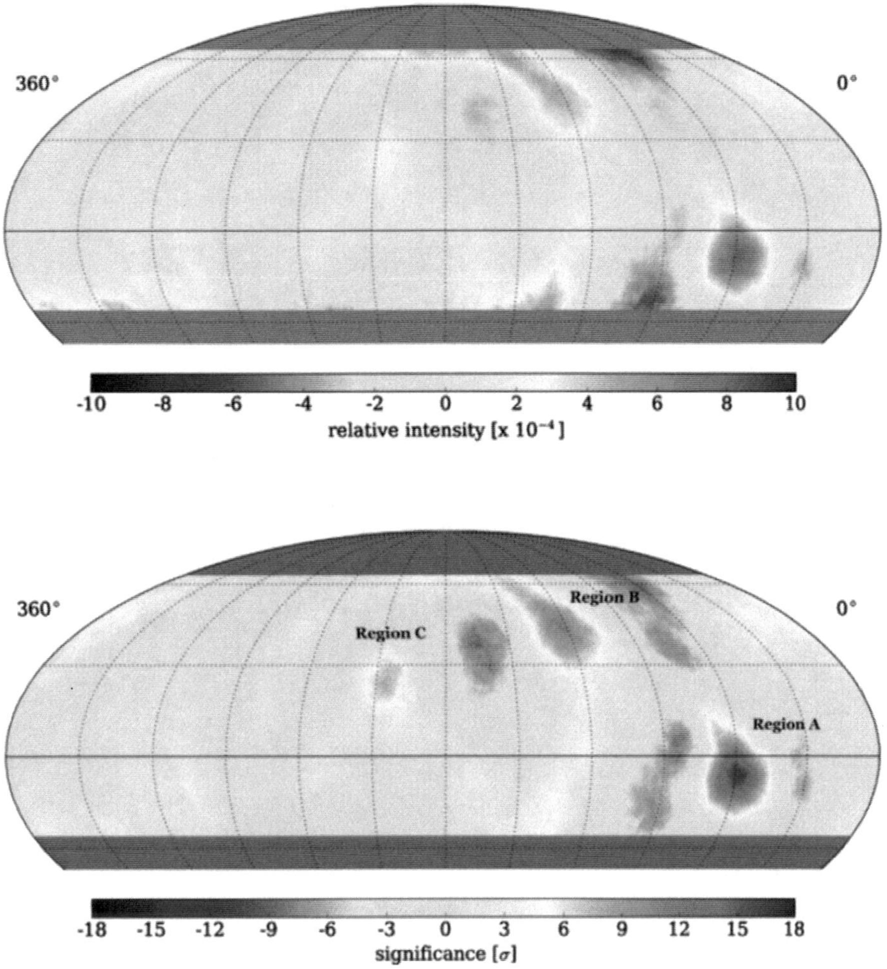

Figure 1. Relative intensity (top) and pre-trial significance (bottom) of the cosmic-ray flux after fit and subtraction of the dipole, quadrupole and octupole term. The coordinates are equatorial. Right ascension runs from 0° to 360° from right to left. The solid horizontal line denotes a declination of 0° [4].

almost continuos monitoring of known gamma-ray sources and the detection of transient events at VHE. These transient events include the possible detection of gamma-ray bursts (GRBs). Extrapolating data from GRB observations by Fermi-LAT, it is expected that HAWC observes 1 or 2 GRBs per year [9]. Another interesting kind of transient events are the high activity states (flares) of the active galactic nuclei (AGNs). This flares by themselves are interesting to study acceleration mechanisms inside the jets of AGNs. These flares usually produce large fluxes from which good quality spectra and skymaps can be produced, and be used to study the interaction of the gamma rays with the extragalactic background light and the presence of intergalactic magnetic fields. The large field of view of HAWC also makes it an optimal instrument for the study of large portions of the sky, searching for extended gamma-ray sources that could be originated by the self-annihilation of dark matter [10].

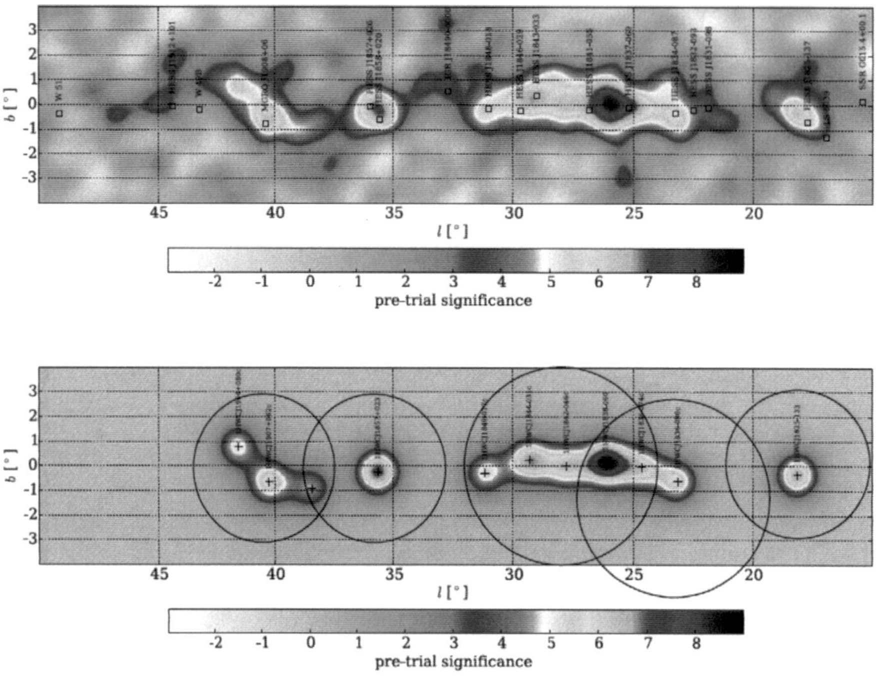

Figure 2. Top: significane map of the inner galactic plane; Bottom: model with 111 seed sources (crosses). The Black circles indicate the five regions of interest [7]. Open squares mark TeV sources in [8]

5. Ongoing HAWC developments

The software and hardware of the HAWC observatory are in continuous development and improvement. Each improvement on the software, once fully tested, is integrated in the on-line analysis of the HAWC observatory. This also means that all the previous data from the observatory is reprocessed. These improvements include changes in the algorithms for the reconstruction of events that improve the angular resolution and increase the sensitivity of the instrument.

On the side of the hardware, a major improvement will be the installation of the outriggers. For the reconstruction of events, it is critical to have a good determination of the "core" of the shower. Most of the large EAS detected by HAWC had their cores falling outside of the detector. To overcome this problem, 300 smaller WCDs (1.6 m diameter, 1.6 m high) will be installed around the main array, equipped with a single PMT each. With this outrigger array it is expected a gain of a factor of 3-4 in sensitivity for gamma rays above 10 TeV over what is presently achieved. The deployment will follow the geometry shown in Fig. 3, in which the inhomogeneities come from the conditions of the terrain.The commissioning of the outrigger array will be in spring 2017 [11].

Acknowledgments

We acknowledge the support from: US National Science Foundation (NSF); US Department of Energy Office of High-Energy Physics; The Laboratory Directed Research and Development (LDRD) program of Los Alamos National Laboratory; Consejo Nacional de Ciencia y Tecnología (CONACyT), Mexico; Red de Física de Altas Energías, Mexico; DGAPA-UNAM; VIEP-BUAP; the University of Wisconsin Alumni Research Foundation; the Institute of Geophysics, Planetary

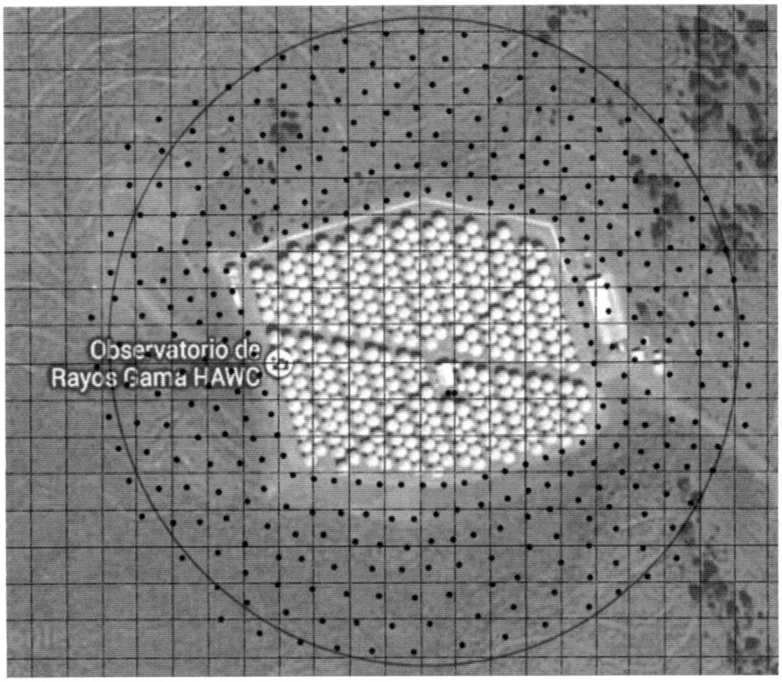

Figure 3. HAWC site with the current HAWC array and the layout of the outrigger array (black points) [11]

Physics, and Signatures at Los Alamos National Laboratory.

References

[1] Atwood W B *et al.* 2009 *Astrophys. J* **697** 1071
[2] Abdo A A *et al.* 2009 *Astrophys. J* **698** 2121
[3] Abeysekara A *et al.* 2013 *Astroparticle Phys.* **50-52** 26
[4] Abeysekara A *et al.* 2014 *Astrophys. J* **796** 108
[5] Amenomori M *et al.* 2005 *Astrophys. J* **626** L29
[6] Bartoli B *et al.* 2013 *Phys. Rev. D* **88** 082001
[7] Abeysekara A *et al.* 2016 *Astrophys. J* **817** 3
[8] Wakely S P and Horan D 2008 *ICRC* **3** 1341
[9] Taboada I and Gilmore R 2013 *Nucl. Instrum. Meth. A* **742** 276
[10] Abeysekara A *et al.* 2014 *Phys. Rev. D* **90** 122002
[11] Sandoval A 2015 *ICRC* arXiv:1509.04269 [astro-ph.IM]

Journal of Physics: Conference Series **730** (2016) 012014 doi:10.1088/1742-6596/730/1/012014

Cluster physics and the importance of forbiddenness

P. O. Hess

Instituto de Ciencias Nucleares, UNAM, Circuito Exterior, C.U., A.P. 70-543, 04510 México D.F., Mexico

E-mail: hess@nucleares.unam.mx

H. Yépez-Martínez

Universidad Autónoma de la Ciudad de México, Prolongación San Isidro 151, Col. San Lorenzo Tezonco, Del. Iztapalapa, 09790 México D.F., Mexico

E-mail: huitzilin_yepez@yahoo.com.mx

Abstract. We review the concept of forbiddenness as introduced by Smirnov and Yu M. Tchuvil'sky [1], which states that most of the properties of nuclear reactions and cluster structure is based on structural effects (Pauli exclusion principle). This becomes also clear when one tries to obtain the ground state of the united nucleus by two heavy clusters, which is not possible when all excitations are put into the relative motion. Smirnov et al. introduced the concept of *forbiddenness*, which is defined as *the minimal number of relative oscillation quanta which have to be shifted to internal excitations of the clusters, such that the ground state can be reached.* We show that Smirnov et al. committed and error in the derivation, leading us to reevaluate the derivation of the forbiddenness. We deduce the correct expression for the forbiddenness, leading to a simple interpretation and implementation in a cluster model and some applications are presented.

1. Introduction

The cluster model in nuclear physics, as introduced by Wildermuth and Kanellopoulos [2, 3], relates the shell model of the united nucleus, with its $SU(3)$ structure for light nuclei, to the shell model of two or more clusters. For heavy clusters, the pseudo-$SU(3)$ model can be applied [4, 5], which takes into account effectively the spin-orbit interaction, allowing still a $SU(3)$ picture. In order to satisfy minimally the *Pauli-exclusion principle*, for example in a two cluster system a *minimal number of oscillation quanta* has to exists in the relative motion. However, still antisymmetrization between the two clusters has to be applied, leading to quite involved calculations. This was resolved in [6, 7], where the $SU(3)$ irreducible representations (irreps) of the clusters were coupled with the irrep of the relative motion and the result finally compared to the complete shell model content. The overlap of the list of the $SU(3)$ coupling with the shell model content gives the microscopic model space of the *Semimicroscopic Alegbraic Cluster Model* (SACM). The SACM was applied with great success to many nuclei. We only mention the study of decay preferences in clusterization of ternary systems [8], hyperdeformed systems [9], energetic considerations in clusterization [8, 10], studies of shape isomers in ^{56}Ni [11] and radiation capture in [12]. In [13] nuclei with the structure core+α where considered, of great

Journal of Physics: Conference Series **730** (2016) 012014 doi:10.1088/1742-6596/730/1/012014

importance in understanding fusion processes in heavy stars. These are only the most recent activities but there are many more in the past within the SACM.

Within the SACM, in order to satisfy the Pauli-exclusion principle, all missing oscillation quanta were put into the relative motion. This procedure works as long as the lighter cluster is not "too heavy". Smirnov et al. [1] demonstrated that when the light cluster is too large (starting from A greater than 12), the ground state of the united nucleus cannot be reached, except if one allows internal excitations in the clusters, i.e., part of the missing oscillation quanta are transferred to the internal excitation of the clusters. *The minimal number of quanta needed to achieve an overlap is called the forbiddenness*, n_c. They also illustrated that fusion and fission properties are governed by the forbiddenness. For example, if in a fusion process for a given system of two colliding nuclei the forbiddenness is large (the ground state can not be reached), the complete fusion is more inhibited than when the forbiddenness is lower.

Thus, in order to describe the ground state of heavy nuclei within the SACM, the forbiddenness has to be determined. Unfortunately, it turned out that the derivation of the forbiddenness in [1] is flawed and we were obliged to obtain it from first principles. This flaw forced others to define an alternative forbiddenness [15], which at least allowed to judge approximately which clusterization is more allowed compared to others, see Ref. [8].

In this contribution be will shortly explain the deduction of the forbiddenness, published elsewhere [14]. In section 2 we will discuss the derivation, showing where the considerations in [1] went wrong. As a result we will obtain a very simple rule on how to obtain the value of the forbiddenness, which also allows to do explicit calculations of heavy cluster systems in future. At the end of the same section two applications are presented and in section 3 conclusions are drawn.

2. Determination of the forbiddenness

In [1] a simple example was considered to illustrate the need of the concept of forbiddenness: ^{40}Ca+^{40}Ca $\to$ ^{80}Zr. The $U(3)$ irreps of these nuclei are $[20,20,20]$ for ^{40}Ca and $[60,60,60]$ for ^{80}Zr, i.e., all $SU(3)$ ireps are $(0,0)$. The number of oscillation quanta in ^{40}Ca is 60, thus the sum of the two clusters is 120. The number of oscillation quanta in ^{80}Zr is 180. The Wildermuth condition states that in this case additional 60 quanta have to be added in order to satisfy the Pauli exclusion principle. If all these quanta are put into the relative motion, as done in the SACM, the $SU(3)$-irrep of the relative motion is $(60,0)$, which has to be coupled with the cluster irrep $(0,0)$, leading to $(\lambda,\mu) = (60,0)$, i.e., *not to the ground state of* ^{80}*Zr* !

The solution is to allow excitations in the cluster system: Let $[n_r,0,0]$ be the irrep of the relative motion (its $SU(3)$-irrep is $(n_r,0)$), in order to couple to the ground state of the Zr-nucleus, we need an $SU(3)$-irrep $(0,n_r)$, which is obtained by the cluster $U(3)$-irrep $[40+n_r,40+n_r,40]$. Thus the number of excitation quanta in the cluster system is $n_c = 2n_r$. When n_0 denoted the minimal number of oscillation quanta to add according to Wildermuth ($n_0 = n_c + n_r$), then $n_0 = 3n_r = 60$, or $n_c = 2n_r = 40$. This number was not obtained in [1].

We thus needed to reanalyze the calculations in order to obtain a new answer.

The coupling of the cluster system, with the $SU(3)$-irrep (λ_c,μ_c), with the relative motion $(n_r,0)$ to a total irrep (λ,μ) can be cast into the following expression of $U(3)$-irreps:

$$\begin{aligned} & [\lambda_c+\mu_c+a,\mu_c+b,c]\otimes[n_r,0,0] \\ \to\ & [\lambda_c+\mu_c+a+k_1,\mu_c+b+k_2,c+k_3] \quad , \end{aligned} \tag{1}$$

where we skipped all possible columns with three boxes in the Young diagram. For example, in the case of ^{40}Ca the $U(3)$ Young diagram is given by $[20,20,20]$, which is equivalent to $[0,0,0]$ and both give the $SU(3)$ irrep (0,0). The cluster irrep is given by $[\lambda_c+\mu_c,\mu_c,0]$ to which we

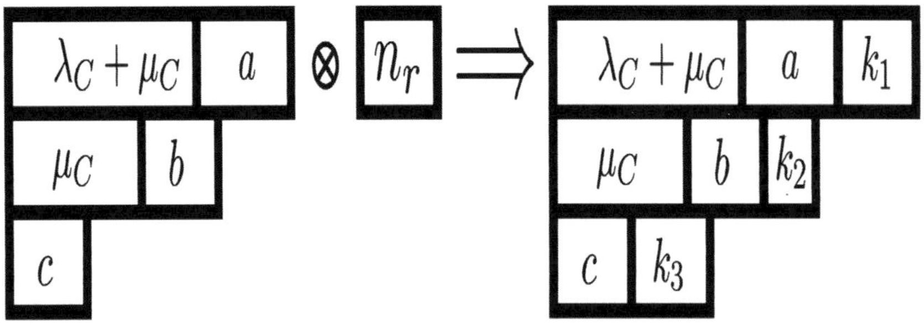

Figure 1. Graphical representation of the direct product of Eq. (1). The $a+b+c$ boxes added to the cluster irrep are distributed within the three rows. The number of quanta from the relative motion $n_c = k_1 + k_2 + k_3$ are also distributed after having applied the direct product. The arrow indicates the coupling to the Young diagrams with the structure shown. In general a sum of many irreps appear.

add a certain number of boxes, the sum given by the excitation quanta $n_c = a+b+c$. a boxes are added in the first row, b in the second and c in the third one. Finally, the relative motion is added such that k_i boxes are added to the i'th row, with $n_r = k_1 + k_2 + k_3$. In all steps, the number of boxes in the first row of a Young diagram has to be larger or equal than those in the second row and these larger or equal than those in the third row. An illustration of the coupling procedure is given in Figure 1. This leads to the following relations:

$$\begin{array}{rcl} 0 & \leq & k_2 \leq \lambda_c + \lambda_0 \; , \; 0 \leq k_3 \leq \mu_c + \mu_0 \\ \lambda & = & \lambda_c + \lambda_0 + k_1 - k_2 \; , \; \mu = \mu_c + \mu_0 + k_2 - k_3 \\ \lambda_0 & = & a - b \; , \; \mu_0 = b - c \quad , \end{array} \tag{2}$$

where λ is given by the difference in the number of boxes of the first row to the second one and μ by the difference in the number of boxes of the second row to the third one.

The relations in (2) have to be resolved with the constraint that n_c is *minimal.* In [14] the explicit derivation is given and here we restrict to the main result:

$$\begin{array}{rcl} n_c & = & \max\left[0, \frac{1}{3}\left\{n_0 - (\lambda - \mu) - (2\lambda_c + \mu_c)\right\}\right] \\ & & +\max\left[0, \frac{1}{3}\left\{n_0 - (\lambda + 2\mu) + (\lambda_c - \mu_c)\right\}\right] \quad . \end{array} \tag{3}$$

The n_0 is given by the Wildermuth condition and (λ, μ) is determined by the ground state or any other excited state of the united nucleus, thus these numbers are fixed. The only numbers which still can be changed are λ_c and μ_c. In order to obtain a minimal value for n_c the $(2\lambda_c + \mu_c)$ has to be *maximized* and $(\lambda_c - \mu_c)$ has to be *minimized.* The first relation implies a large deformed system, while the last one requires that this system is as oblate as possible ($\mu_c > \lambda_c$), which can be interpreted as a large most compact, oblate system.

To determine the forbiddenness is now straight forward. For example, for the case, given above, we have $n_0 = 60$ and $(\lambda, \mu) = (0,0)$ and $(\lambda_c, \mu_c) = (0,0)$. Thus, $n_c = \max\left[0, \frac{60}{3}\right] + \max\left[0, \frac{60}{3}\right] = \frac{120}{3} = 40$.

Table 1. A series of 2-cluster systems, all belonging to the nucleus $^{236}_{92}U_{144}$ are enumerated (first column). In the last column the *forbiddenness* n_c for these systems is evaluated.

No.	Two cluster system	United cluster	n_c
1	$^{4}_{2}He_2 + ^{232}_{90}Th_{142}$	$^{236}_{92}U_{144}$	0
2	$^{20}_{10}Ne_{10} + ^{216}_{82}Pb_{134}$	$^{236}_{92}U_{144}$	4
3	$^{24}_{10}Ne_{10} + ^{212}_{82}Pb_{130}$	$^{236}_{92}U_{144}$	4
4	$^{26}_{10}Ne_{16} + ^{210}_{82}Pb_{128}$	$^{236}_{92}U_{144}$	2
5	$^{28}_{12}Mg_{16} + ^{208}_{80}Hg_{128}$	$^{236}_{92}U_{144}$	4
6	$^{30}_{12}Mg_{18} + ^{206}_{80}Hg_{126}$	$^{236}_{92}U_{144}$	4
7	$^{32}_{14}Si_{18} + ^{204}_{78}Pt_{126}$	$^{236}_{92}U_{144}$	0
8	$^{34}_{14}Si_{20} + ^{202}_{78}Pt_{124}$	$^{236}_{92}U_{144}$	6
9	$^{40}_{22}Ti_{18} + ^{296}_{70}Yb_{126}$	$^{236}_{92}U_{144}$	16
10	$^{66}_{36}Kr_{30} + ^{170}_{56}Ba_{114}$	$^{236}_{92}U_{144}$	20
11	$^{66}_{22}Ti_{44} + ^{170}_{70}Yb_{100}$	$^{236}_{92}U_{144}$	32
12	$^{128}_{50}Sn_{78} + ^{108}_{42}Mo_{66}$	$^{236}_{92}U_{144}$	28
13	$^{132}_{50}Sn_{82} + ^{104}_{42}Mo_{62}$	$^{236}_{92}U_{144}$	36

The forbiddenness will play an important role in extending the SACM, allowing that the light clusters have a large mass and the treatment of heavy nuclei. In the Hamiltonian of the model a dependence of n_c has to be included and also for the calculation of the spectroscopic factor. In [16] an algebraic expression for the spectroscopic factor was given, which is able to reproduce for cluster+α systems the within the $SU(3)$ model theoretically determined spectroscopic factors of light nuclei [16] by 1 to 3 percent. The formula depends on (λ_i, μ_i) (i=1,2), the $SU(3)$ irreps of the clusters in their ground state, (λ_c, μ_c), the cluster irrep, (λ, μ), the final $SU(3)$ irrep and $n_\pi = n_0 + \Delta n_\pi$, the number of quanta in the relative motion. In the new expression, (λ_c, μ_c) is changed to $(\lambda_c + \lambda_0, \mu_c + \mu_0)$ and n_π to $n_0 - n_c + \Delta n_\pi$, where Δn_π is the number of relative excitation quanta with respect to $0\hbar\omega$.

In Table 1 the forbiddenness is listed for different partitions of ^{236}U. When the light cluster has $A < 12$, the forbiddenness is zero and only in rare example it might be zero for larger clusters. In figure 2 the forbiddenness is plotted versus the mass of the light cluster. As can be appreciated, the forbiddenness increases toward larger clusters, which reflects the property that in order the system can fission into the two clusters, both clusters have to be in general in an excited state. The inverse process, i.e. fusion, also is largely inhibited allowing molecular states, but reducing the probability for fusion.

The same data are listed in Table 2 for ^{252}Cf and the forbiddenness is plotted versus the mass of the light cluster in Figure 3.

In both systems, we see the structural effect (the Pauli principle) on the structure of fission and fusion processes, as discussed in [1]. The implication of the extension of the SACM is clear: Not only we have to include in the Hamiltonian a dependence on the forbiddenness but also the algebraic expression of the spectroscopic factor, as proposed in [16].

3. Conclusions

We have derived an expression for the forbiddenness, as defined in [1]. The result is easy to implement within the SACM, for the inclusion of systems where the light cluster has a large mass. Using the pseudo-$SU(3)$ model [4, 5], this concept can also be introduced easily for heavy nuclei, to which we will dedicate our effort in near future.

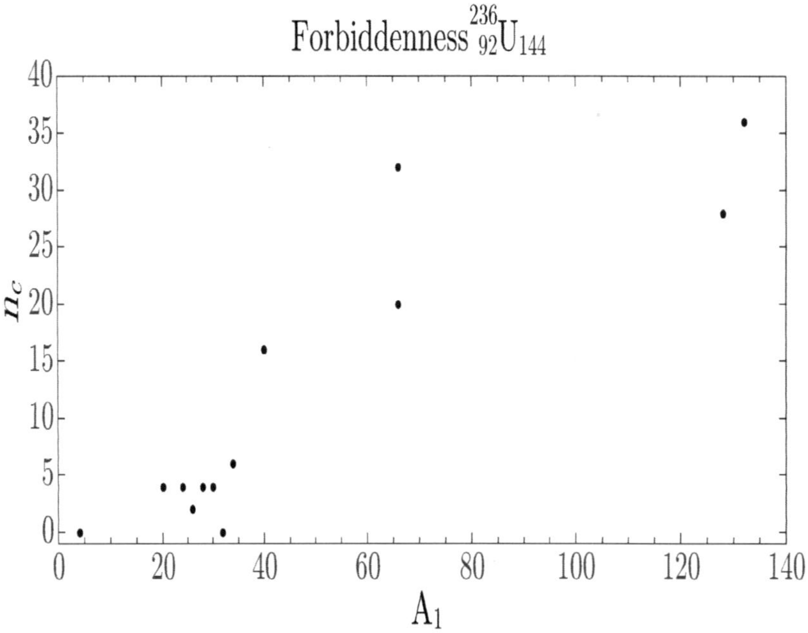

Figure 2. Graphical representation of the results from Table 1. The forbiddenness is depicted versus the mass of the lightest cluster.

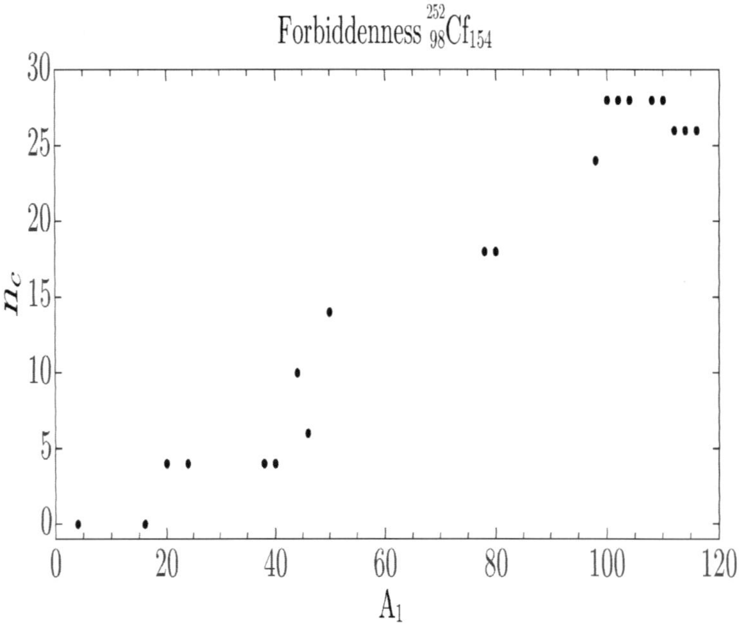

Figure 3. Graphical representation of the results from Table 2. The forbiddenness is depicted versus the mass of the lightest cluster.

Table 2. A series of 2-cluster systems, all belonging to the nucleus $^{252}_{98}\mathrm{Cf}_{154}$, $(\lambda, \mu) = (56, 10)$, are enumerated (first column). In the last column the *forbiddenness* n_c for these systems is evaluated.

No.	Two cluster system	United cluster	n_c
1	$^{4}_{2}He_2 + ^{248}_{96}Cm_{152}$	$^{252}_{98}Cf_{154}$	0
2	$^{16}_{8}O_8 + ^{236}_{90}Th_{146}$	$^{252}_{98}Cf_{154}$	0
3	$^{20}_{6}C_{14} + ^{232}_{92}U_{140}$	$^{252}_{98}Cf_{154}$	4
4	$^{24}_{10}Ne_{14} + ^{228}_{88}Ra_{140}$	$^{252}_{98}Cf_{154}$	4
5	$^{38}_{14}Si_{24} + ^{214}_{84}Po_{130}$	$^{252}_{98}Cf_{154}$	4
6	$^{40}_{16}Sn_{24} + ^{212}_{82}Pb_{130}$	$^{252}_{98}Cf_{154}$	4
7	$^{44}_{16}Sn_{28} + ^{208}_{82}Pb_{126}$	$^{252}_{98}Cf_{154}$	10
8	$^{46}_{18}Ar_{28} + ^{206}_{80}Hg_{126}$	$^{252}_{98}Cf_{154}$	6
9	$^{50}_{18}Ar_{32} + ^{206}_{80}Hg_{122}$	$^{252}_{98}Cf_{154}$	14
10	$^{78}_{30}Zn_{48} + ^{174}_{68}Er_{106}$	$^{252}_{98}Cf_{154}$	18
11	$^{80}_{30}Zn_{50} + ^{172}_{68}Er_{104}$	$^{252}_{98}Cf_{154}$	18
12	$^{98}_{38}Sr_{60} + ^{152}_{60}Nd_{92}$	$^{252}_{98}Cf_{154}$	24
13	$^{100}_{38}Sr_{62} + ^{152}_{60}Nd_{92}$	$^{252}_{98}Cf_{154}$	28
14	$^{100}_{40}Zr_{60} + ^{152}_{58}Ce_{94}$	$^{252}_{98}Cf_{154}$	28
15	$^{102}_{40}Zr_{62} + ^{150}_{58}Ce_{92}$	$^{252}_{98}Cf_{154}$	28
16	$^{104}_{40}Zr_{64} + ^{148}_{58}Ce_{90}$	$^{252}_{98}Cf_{154}$	28
17	$^{104}_{42}Mo_{62} + ^{148}_{56}Ba_{92}$	$^{252}_{98}Cf_{154}$	28
18	$^{108}_{42}Mo_{66} + ^{144}_{56}Ba_{88}$	$^{252}_{98}Cf_{154}$	28
19	$^{110}_{44}Ru_{66} + ^{142}_{54}Xe_{88}$	$^{252}_{98}Cf_{154}$	28
20	$^{112}_{44}Ru_{68} + ^{140}_{54}Xe_{86}$	$^{252}_{98}Cf_{154}$	26
21	$^{114}_{44}Ru_{70} + ^{138}_{54}Xe_{84}$	$^{252}_{98}Cf_{154}$	26
22	$^{116}_{46}Pd_{70} + ^{136}_{52}Te_{84}$	$^{252}_{98}Cf_{154}$	26

Acknowledgments
The authors acknowledge financial help from DGAPA-PAPIIT (IN100315).

References

[1] Smirnov Yu F and Tchuvil'sky Yu M 1982 in *XXXIIth Conference on Nuclear spectroscopy and Nuclear Structure* (Leningrad) 231
[2] Wildermuth K and Kanellopoulos Th 1958 *Nucl. Phys.* **7** 150
[3] Wildermuth K and Tang Y C 1977 *A Unified Theory of the Nucleus* (New York, Academic Press)
[4] Hecht K T and Adler A 1969 *Nucl. Phys. A* **137** 129
[5] Arima A, Harvey M and Shimizu K 1969 *Phys. Lett. B* **30** 517
[6] Cseh J 1992 *Phys. Lett. B* **281** 173
[7] Cseh J and Lévai G 1994 *Ann. Phys. (N.Y.)* **230** 165
[8] Algora A, Cseh J, Darai J and Hess P O 2006 *Phys. Lett. B* **639** 451
[9] Cseh J *et al.* 2010 *J. Phys.: Conf. Ser.* **239** 012006
[10] Darai J, Cseh J and Jenkins D J 2012 *Phys. Rev. C* **86** 064309
[11] Darai J *et al.* 2011 *Phys. Rev. C* **84** 024302
[12] Lebhertz D *et al.* 2012 *Phys. Rev. C* **85** 034333
[13] Yépez-Martínez H, Ermamatov M J, Fraser P R and Hess P O 2012 *Phys. Rev. C* **86** 034309
[14] Yépez-Martínez H and Hess P O 2015 *J. Phys. G* **42** 095109
[15] Algora A and Cseh J 1996 *J. Phys. G* **22** L39
[16] Hess P O, Algora A, Cseh J and Draayer J P 2004 *Phys. Rev. C* **70** 051303(R)
[17] Gupta R and Greiner W 1994 *Int. J. Mod. Phys. E* **3** 335

Journal of Physics: Conference Series **730** (2016) 012015 doi:10.1088/1742-6596/730/1/012015

Electromagnetic Properties of the $I^\pi = 11/2^-$ isomers in the cadmium isotopes: What can we learn from them?

Y. Lei[1], S. Pittel[2] and H. Jiang[3,4]

[1] Key laboratory of neutron physics, Institute of Nuclear Physics and Chemistry, China Academy of Engineering Physics, Mianyang 621900, China
[2] Bartol Research Institute and Department of Physics and Astronomy, University of Delaware, Newark, Delaware 19716, USA
[3] School of Arts and Sciences, Shanghai Maritime University, Shanghai 201306, China
[4] Department of Physics, Shanghai Jiao Tong University, Shanghai 200240, China

E-mail: `pittel@bartol.udel.edu`

Abstract. The electromagnetic features of the $11/2^-$ isomers in the odd-mass Cd isotopes are shown to exhibit an an anomaly near $N = 70$. We report shell-model calculations of these isotopes aimed at describing these properties. We find a sudden phase change in the hexadecapole component of the wave functions precisely at $N = 70$, which gives rise to different linear relations between the Q and μ values before and after $N = 70$, as needed to reproduce the experimental data. The particle-hole transformation properties associated with a strong subshell closure at $N = 70$ followed by fairly close-lying neutron orbitals from $N = 70 - 82$ is suggested as a possible explanation for this phase change.

1. Introduction

A recent study of the $I^\pi = 11/2^-$ isomers in the Cd-isotope chain [1] indicates that the quadrupole and magnetic moments (denoted by Q and μ, respectively) both vary roughly as linear functions of neutron number. However, as was also noted, there is a noticeable difference in the linear relation for the μ values between $^{111-117}$Cd and $^{119-127}$Cd, separated by $N = 70$. Indeed, if one performs a linear fit to the quadrupole moments, as shown in fig. 1(a), a similar behavior is observed: the best-fit linear relation perfectly describes the quadrupol moments of $^{111-117}$Cd, while residues from the same fit obviously exceed experimental errors for $^{119-127}$Cd, as shown in fig. 1(b). Thus, the quadrupole moment data also exhibit a change in linear behavior before and after $N = 70$.

A density functional study [2] recently reported suggested that the Q linearity in the Cd isotopes has a different mechanism for $N < 70$ and $N > 70$. In this work, we study the mechanism for the behavior of both the quadrupole moments and magnetic moments at $N = 70$ in the framework of a simple shell-model treatment, to see whether we can explain the data and in doing so obtain new insight on the structural properties of these isotopes. Further details can be found in [3] where the results of this study were first presented.

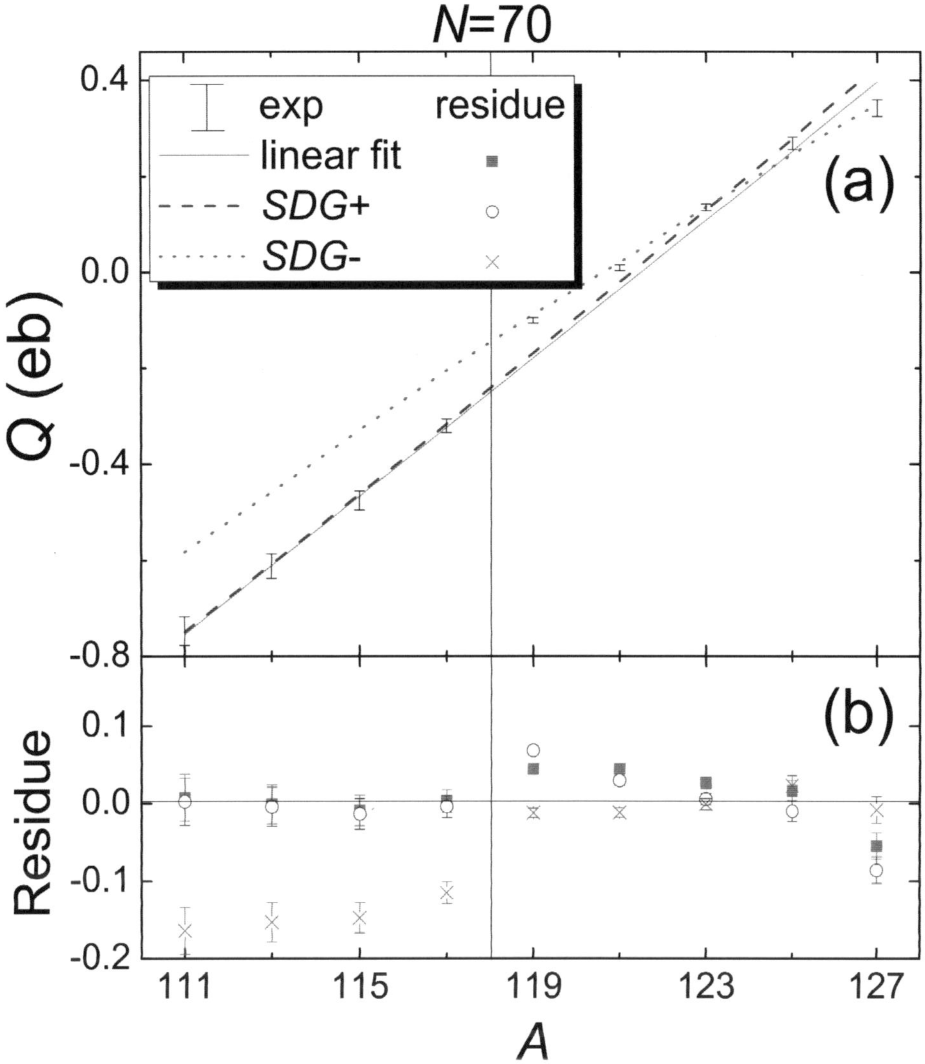

Figure 1. Quadrupole moments of the $11/2^-$ isomers in the Cd isotopes. In panel (a), we compare experimental values [1], linear fitting results, and results from the $SDG+$ and $SDG-$ calculation. Panel (b) presents residues of the linear fit. The $N = 70$ position is highlighted. For $N < 70$, the residues are smaller than the experimental errors, so that a linear relation follows. In contrast, the residues for $N > 70$ are larger than experimental errors, which means that the same linear behavior is not respected. The $SDG+$ and $SDG-$ results are consistent with the $N < 70$ and $N > 70$ experimental data, respectively, as discussed in the text.

2. Calculations and Analysis

We begin by outlining the basic ingredients of our calculations:

1 We first assume simple, but reasonable, neutron and proton configurations from which to build the $11/2^-$ isomers.

2 We then calculate the quadrupole and magnetic-dipole matrix elements for the above configurations.

3 We next consider several trial wave functions for the isomers and fit their respective amplitudes to optimally describe the quadrupole- and magnetic-moment data for the $11/2^-$

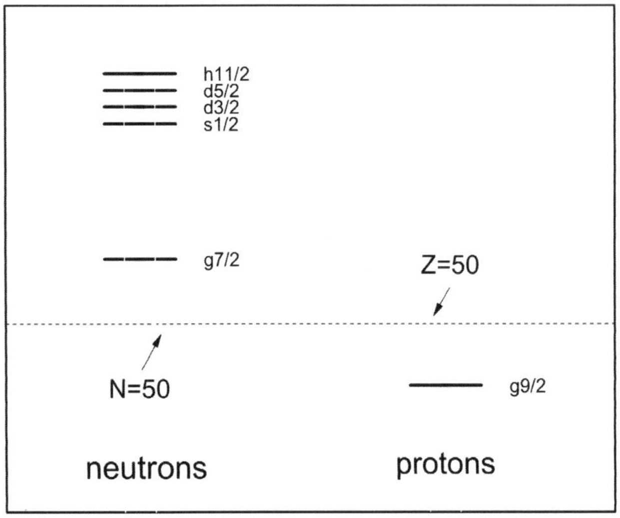

Figure 2. Schematic illustration of the single-particle levels of relevance to the Cd isotopes and their corresponding energies.

isomers.

As we will see, the procedure leads to a set of wave functions that nicely describe the evolution of the quadrupole and magnetic moments of the $11/2^-$ isomers, including the change in behavior at $N = 70$, and furthermore suggests a fairly simple picture as to why it occurs.

In fig. 2, we schematically show the relevant single-particle energy levels appropriate to our analysis of the $N = 48$ Cd isotopes. For protons, only the $g_{9/2}$ orbital (for holes) is relevant. For neutrons, the levels of the $N = 50 - 82$ shell are all in principle relevant. However, the $g_{7/2}$ orbital is fairly well isolated from the others, which are themselves fairly close in energy.

With that as background, we now describe the model space we use in our shell-model study. For the proton configuration, we assume that the two holes in the $Z = 50$ core both lie in the $g_{9/2}$ orbital. There are five such two-hole states, with total spins $J = 0, 2, 4, 6, 8$, which we denote $S_\pi = (g_{9/2} \otimes g_{9/2})^0$, $D_\pi = (g_{9/2} \otimes g_{9/2})^2$, $G_\pi = (g_{9/2} \otimes g_{9/2})^4$, $I_\pi = (g_{9/2} \otimes g_{9/2})^6$, and $K_\pi = (g_{9/2} \otimes g_{9/2})^8$, respectively.

Since the Cd isotopes are nearly spherical, we assume minimal proton-neutron configuration mixing. We thus assume, following [1], the simplest possible configuration for neutrons, a seniority-one configuration involving a single unpaired $h_{11/2}$ particle and the remaining particles paired to spin-zero and distributed over the close-lying levels shown in fig. 2. We denote this neutron configuration as ϕ_ν.

With the above assumptions, our basis space for treating the $11/2^-$ isomers consists of only five configurations: $\tilde{S} = \phi_\nu \otimes S_\pi$, $\tilde{D} = \phi_\nu \otimes D_\pi$, $\tilde{G} = \phi_\nu \otimes G_\pi$, $\tilde{I} = \phi_\nu \otimes I_\pi$, and $\tilde{K} = \phi_\nu \otimes K_\pi$.

The Q and μ matrix elements for the neutron configurations can be calculated straightforwardly based on the seniority-one nature of ϕ_ν, using the well-known seniority

Journal of Physics: Conference Series **730** (2016) 012015 doi:10.1088/1742-6596/730/1/012015

reduction formulae [4] that apply for either a single orbit or a set of *degenerate* orbits, as assumed to approximately apply here. The proton Q and μ matrix elements of the $(g_{9/2})^{-2}$ configurations can likewise be calculated simply based on the same formalism.

We need to choose appropriate effective charges and Lande g factors for the analysis. The proton effective charge are taken from a fit to the experimental quadrupole moment of the 2_1^+ state of ^{112}Cd, which should have as predominantly D_π structure. Some minor modifications to this simple prescription are described in [3]. The usual Lande g factors are assumed. The neutron effective charge will be obtained from a fit to the quadrupole moments, to be described below.

Since S_π and D_π configurations dominate the low-lying states of the Cd isotopes, we first consider as a trial wave-function $C_S\tilde{S} + C_D\tilde{D}$, where C_S and C_D are wave function amplitudes. The corresponding probabilities C_S^2 and C_D^2 can be obtained from the magnetic moment data on the $11/2^-$ isomers. Using the values that emerge, we then determine the neutron effective charge e_ν so as to optimize the reproduction of the associated quadrupole moments. We consider both possibilities for the relative sign of C_D with respect to C_S, with the eventual choice determined by which yields results in closer accord to experiment. This is referred to as our "*SD* fit". We present the optimal e_ν and the root-mean-square deviation (RMSD) of the *SD* fit in the top row of Table 1. Both the resulting $e_\nu = 7.334$e and the RMSD=0.456eb are much too large for this to be viewed as an acceptable fit. Thus we need to consider additional proton configuration.

Table 1. Fitting results with *SD*, *SDG*, *SDI* and *SDK* trial wave functions.

	e_ν (e)	RMSD (eb)
SD	7.334±2.110	0.456
SDG	1.496±0.221	0.009
SDI	3.079±6.479	0.018
SDK	3.073±10.44	0.018

We thus consider three other trial wave-functions: (a) $C_S\tilde{S}+C_D\tilde{D}+C_G\tilde{G}$, (b) $C_S\tilde{S}+C_D\tilde{D}+C_I\tilde{I}$ and (c) $C_S\tilde{S} + C_D\tilde{D} + C_K\tilde{K}$, which we denote "*SDG*, *SDI* and *SDK*", respectively. Here too C_G, C_I and C_K are wave function amplitudes to be determined analogously through optimization. The final best-fit values and the RMSD are also listed in Table 1. All three fits achieve reasonable levels of agreement, with RMSD values$\sim$ 0.01eb. Therefore, in what follows we limit ourselves to these trial wave functions.

When comparing the various fits, we conclude that the *SDG* fit seems to be the best for two principal reasons:

- The *SDG* fit provides the smallest RMSD.
- The *SDI* and *SDK* fits both yield an unusually large $e_\nu \sim$ 3e value, as did the *SD* fit discussed earlier.

Therefore, we only consider the *SDG* wave-function and their associated best-fit e_ν value in the analysis to follow.

For the *SDG* wave functions, the choice of the C_G phase relative to that of C_S can lead to different quadrupole moments. We classify the *SDG* wave functions according to the relative phase of C_G with respect to C_S (assumed positive), with *SDG*+ having $C_G > 0$ and and *SDG*− having $C_G < 0$.

In fig. 1, the calculated quadrupole moments associated with the *SDG*+ and *SDG*− wave functions are compared with the experimental quadrupole moments. The *SDG*+ results are

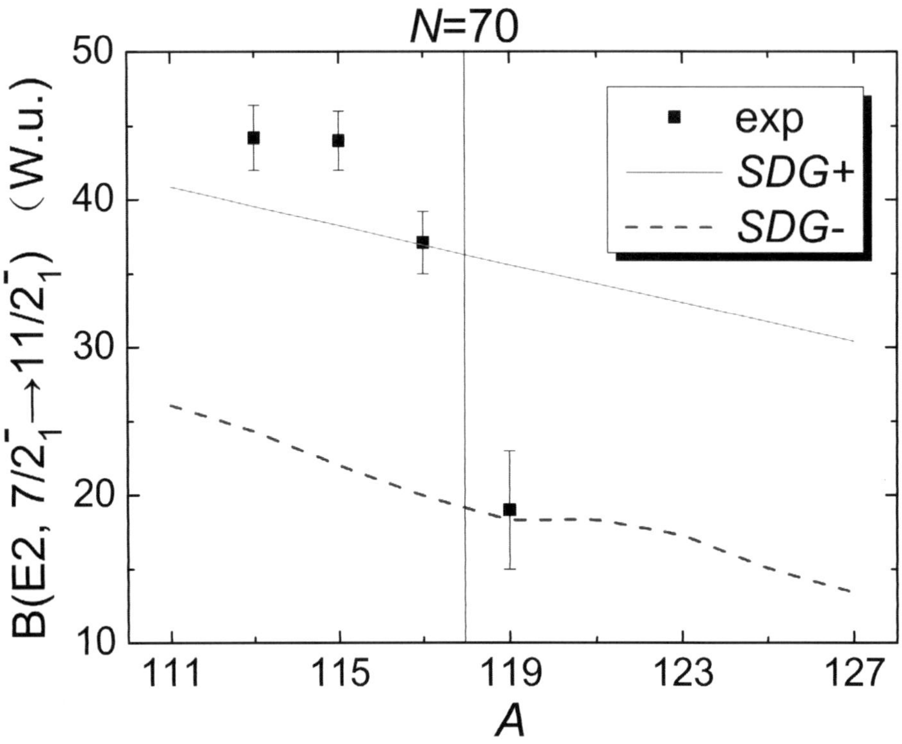

Figure 3. Calculated and experimental B(E2, $7/2^-_1 \rightarrow 11/2^-_1$) values. Calculated results based on both the $SDG+$ and $SDG-$ wave functions for the $11/2^-$ isomers are presented. The experimental data (exp) is from ref. [5]. The sudden $N = 70$ phase change is highlighted.

compatible with the best-fit linear relation of the experimental quadrupole moments, and provide a better description for $N < 70$. In contrast, the $SDG-$ results exhibit some curvature beyond $N = 70$, as in the experimental data. From these results, we conclude that there is a C_G phase change across $N = 70$. This provides a plausible explanation for the change in Q linearity on the two sides of $N = 70$, as reflected in fig. 1.

The precise difference between the $SDG+$ and $SDG-$ quadrupole moments is observable but not very pronounced. It is desirable therefore to use other electromagnetic data to more clearly isolate the C_G phase in the various isotopes. In particular, we will consider the systematic strong B(E2, $7/2^-_1 \rightarrow 11/2^-_1$) values in $^{113-119}$Cd [5], to see whether they can provide additional information. We will calculate these B(E2) values using both the $SDG+$ and $SDG-$ wave-functions obtained above, and carry out an isotope-by-isotope comparison between the experimental and calculated results.

As necessary input to the B(E2) calculation, we must first identify an appropriate wave function to use for the initial $7/2^-_1$ state. The strong E2 transition $7/2^-_1 \rightarrow 11/2^-_1$ suggests a similarity between the $7/2^-_1$ state and the $11/2^-$ isomeric state We thus treat the $7/2^-_1$ state as a simple proton-neutron recoupling of the $11/2^-$ isomer, namely as $\phi_\nu \otimes D_\pi$ (this should be lower in energy than $\phi_\nu \otimes G_\pi$ and thus more important) *but coupled of course to* $J^\pi = 7/2^-$. Thus, in the spirit of the analysis to date, we assume a single configuration for the state. But of course we still need to see whether a good description of the data follows from such a simple choice.

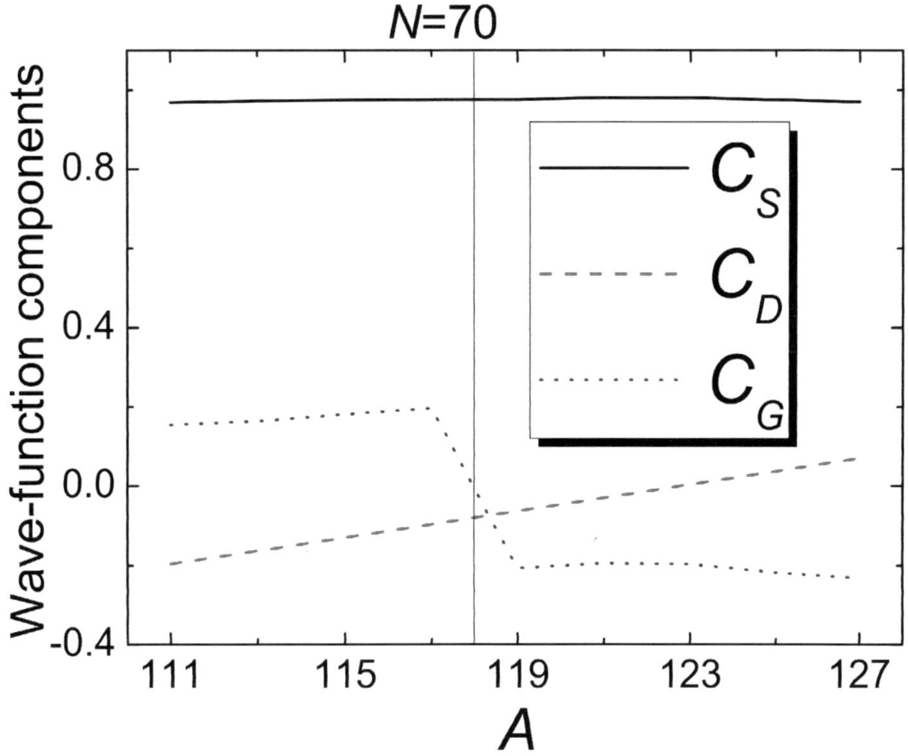

Figure 4. Wave functions from the final SDG fit to the quadrupole moments and B(E2) values. The $N = 70$ phase change in C_G is highlighted.

The calculated B(E2, $7/2_1^- \rightarrow 11/2_1^-$) values that emerge from both the $SDG+$ and $SDG-$ wave functions are compared with the available experimental data in fig. 3. This comparison confirms a sudden phase change of C_G at $N = 70$: the $N < 70$ B(E2) values clearly favor the $SDG+$ results ($C_G > 0$), whereas beyond $N = 70$, only the $SDG-$ calculation can fit the single experimental B(E2) data point. Based on these results, we summarize our final SDG wave functions from the Q-fitting procedure in fig. 4.

What we see from the figure is that the C_S value is close to 1 for all the Cd isotopes. Thus, the $11/2^-$ isomer is constructed with a dominant $h_{11/2}$ single-particle configuration, with perturbations arising from D_π and G_π excitations. The second feature to note is the rapid change of phase exhibited by C_G, precisely at $N = 70$. Lastly, note that that the D_π contributions also changes phase near $N = 70$, but it does so very smoothly.

How might we understand these results? In perturbation theory, changes in C_D and C_G relative to C_S reflect sign changes in the associated $p - n$ interaction coupling matrix elements. The relevant interactions correspond to scalar products of multipole operators in the $Q \cdot Q$ channel (responsible for C_D) and the $H \cdot H$ channel (responsible for C_G) . Whereas the proton reduced matrix elements are simple and do not evolve with neutron number, the neutron reduced matrix elements do evolve with neutron number, and must therefore be responsible for the sign change. As is well known [6], the matrix elements of an even multipole operator, in either a single orbit or a series of degenerate orbits, changes sign at mid-shell due to a change from particle-like to hole-like behavior.

As noted earlier, the neutron orbits from $N = 58-82$ (the $s_{1/2}$, $d_{3/2}$, $d_{5/2}$, and $h_{11/2}$) are fairly close in energy and well isolated from the lower $g_{7/2}$. Thus, they seem to provide the needed well-isolated subshell. We would expect, therefore, that both the quadrupole and hexadecapole matrix elements change sign at mid-shell, which is precisely at $N = 70$. Since C_D evolves very smoothly with N, the effect of the sign change shows up most dramatically in C_G.

On the basis of these arguments, we conclude that the dramatic change in the sign of C_G *seems to be a manifestation of a fairly isolated* $N = 58 - 82$ *subshell with relatively close-lying orbits and the particle-hole transformation properties that arise in such circumstances.*

3. Summary

Summarizing, we have extracted shell-model wave functions of the $11/2^-$ isomers in the odd-mass $^{111-129}$Cd by fitting their electromagnetic features, including both electromagnetic moments and B(E2) values. A phase change in the hexadecapole component of the extracted wave functions explains the change in linear behavior of the quadrupole and magnetic moments for $N < 70$ and $N > 70$ as well as the BE2($7/2_1^- \rightarrow 11/2_1^-$) behavior. According to perturbation theory, this phase change at $N = 70$ is related to a change in sign of the hexadecapole proton-neutron matrix elements as we pass through $N = 70$, if we assume a strong subshell effect taking place for this neutron number.

Acknowledgments

This work was supported in part by the US National Science Foundation under grant # 0553127 and also by the National Natural Science Foundation of China under contracts # 11305151, 11305101, 11247241. One of the authors (J. H.) thanks the Shanghai Key Laboratory of Particle Physics and Cosmology for financial support under grant # 11DZ2260700.

References

[1] Yordanov D T *et al.* 2013 *Phys. Rev. Lett.* **110** 192501
[2] Zhao P W, Zhang S Q, and Meng J 2014 *Phys. Rev. C* **89** 011301
[3] Yang L, Jiang H, and Pittel S 2015 *Phys. Rev. C* **92** 024321
[4] De-Shalit A and Talmi I 1963 *Nuclear Shell Theory* (New York, Academic Press)
[5] Evaluated Nuclear Structure Data File (ENSDF) www.nndc.bnl.gov/ensdf
[6] Lawson R D 1980 *Theory of the Nuclear Shell Model* (New York, Oxford University Press)

Journal of Physics: Conference Series **730** (2016) 012016 doi:10.1088/1742-6596/730/1/012016

Spectroscopy of Light Nuclei with Low Energy Nuclear Reactions

I. Lombardo[1,2], D. Dell'Aquila[1,2] and M. Vigilante[1,2]

[1] Dipartimento di Fisica "E. Pancini", Università di Napoli Federico II, Via Cinthia, 80126, Napoli, Italy

[2] INFN - Sezione di Napoli, Via Cinthia, 80126, Napoli, Italy

E-mail: ivlombardo@na.infn.it

Abstract. We discuss new results concerning the investigation of the ^{19}F(p,α_0)^{16}O and ^{10}B(p,α_0)^{7}Be reactions at low energies. Both reactions are important for the nuclear spectroscopy of the formed compound nucleus, i.e. ^{20}Ne and ^{11}C respectively, and play a role in nuclear astrophysics. For the ^{10}B(p,α_0)^{7}Be case, a comprehensive analysis of our reaction data and other scattering data points out the possible presence of an unreported state in ^{11}C at $E_x \approx 9.36$ MeV. For the ^{19}F(p,α_0)^{16}O case, the study of the low energy angular distributions testifies the role played by low energy resonances in the S-factor, leading to an enhanced reaction rate at stellar energies.

1. Introduction

Nuclear collisions between light ions represent a very useful tool to explore the structure of light nuclei. In particular, it is possible to deduce information about the appearance of α-cluster states in light nuclei. Different types of experimental nuclear techniques can be used to investigate these aspects. For example, direct α-transfer [1, 2, 3] and sequential break-up [4, 5, 6, 7] reactions have been widely used to make the spectroscopy of light nuclei with medium-energy stable and radioactive beams. At low energies, other powerful tools to do nuclear spectroscopy of light nuclei are the study of resonant elastic scattering of ^{4}He nuclei on various targets, both in direct and inverse kinematics [8, 9, 10, 11] or, in general, the study of nuclear reactions leading to the formation of a compound nucleus that decades by α-particle emission [12, 13, 14]. Beyond ion collisions, other techniques, such as radioactive decay [15], electron scattering [16], γ-induced reactions [17] have been used advantageously to investigate the cluster structure of light nuclei.

In the first part of these proceedings we will describe some results on the structure of the ^{11}C nucleus, obtained by analysing the ^{10}B(p,α_0)^{7}Be reaction at 0.6 - 1.1 MeV bombarding energies. Because of various experimental difficulties due to the reaction kinematics and the presence of contaminants in the target, the data reported in the literature in this energy range are very few, and often affected by large uncertainties [18, 19]. Moreover, the spectroscopy of ^{11}C is full of ambiguities in the $E_x = 9.0 - 12$ MeV range, including the energy range discussed here. In our experiment, we measured the integrated reaction cross section and S-factor. They are about 30% lower than previous estimate [20]. An R-matrix fit [21] of ^{10}B(p,α_0)^{7}Be and ^{10}B(p,α_1)^{7}Be reactions and ^{10}B(p,p_0)^{10}B scattering data allowed to determine resonance parameters for excited states in ^{11}C in the excitation energy domain $E_x \approx$9-11 MeV, and pointed out the possible existence of a state at $E_x \simeq$9.36 MeV.

In the second part of these proceedings we will discuss the results obtained by analysing the structure of ^{20}Ne by means of the ^{19}F(p,α_0) reaction at sub-Coulomb energies. The analysis of excitation functions and angular distributions allow to refine the ^{20}Ne spectroscopy in the $E_x \approx$ 13.0-14.0 MeV region [22, 23]. We measured the S-factor distribution in the $E_{cm} \simeq$ 0.18-1.1 MeV range, where very few data have been reported in the literature [24, 25, 26]. The lowermost energy data that we have obtained are close to the stellar energy domain [26]; they are important because this reaction is involved in the CNOF cycle during the hydrogen-burning phase and represents an important channel of fluorine destruction in hydrogen-rich environments of massive stars [27]. Our data point out the presence of low-energy resonances, in agreement with results obtained with indirect experiments based on the Trojan Horse Method [28, 29]. They can increase up to a factor ≈2 the reaction rate at temperatures typical of AGB stars.

2. The ^{10}B(p,α_0)^{7}Be reaction and the structure of ^{11}C

The ^{10}B(p,α)^{7}Be reaction is important in various fields of nuclear physics. In fact, it allows to study the spectroscopy of the non self-conjugated ^{11}C compound nucleus [30], with particular regard to the appearance of α-cluster states (see, e.g. [12, 13]). Furthermore, it is important to accurately describe the destruction of the ^{10}B isotope in stars [19]. Finally it is involved in radio-protection issues in Nuclear Fusion technology [31].

Unfortunately, direct data reported in the literature at low bombarding energies, especially in the $E \approx 1-2$ MeV region, are quite few [18, 20], being often characterized by large error bars. Considering all these reasons, we decided to perform a new experiment to measure the ^{10}B(p,α)^{7}Be reaction in the E_p=1.028-0.630 MeV bombarding energy domain.

Proton beams at bombarding energies E_p=0.630-1.028 MeV were delivered by the TTT3 tandem accelerator of the Laboratorio dell'Acceleratore (LdA) at Federico II University of Naples [32]. The energy step here used was about 40 keV. To reduce pile-up effects, in this experiment the beam intensity did not overcome 1 nA. A boron foil (38 μg/cm^2 thick) was used as reaction target. The target surface was orthogonal to the beam direction. Because of the manufacturing process, the target includes several contaminants (mainly Li, C, O, Al, Cl, Cu, Ba). Due to kinematic and energy loss reasons, peaks due to the heavier of these contaminants overlap with the α_0 signal at backward angles. For this reason, to measure these yields we used a method based on the simultaneous measurements of two ejectile spectra at the same angle (in opposite directions with respect to the beam axis), the first one taken with an unshielded silicon detector, the second one taken by a silicon detector covered by a thin Al foil (3μm thick). This foil is able to stop (or highly slow down) the α_0 particles, allowing the elastically scattered protons to punch through. Then, the α_0 yields have been derived by subtraction, taking into account the energy loss and straggling effects. Details on this experimental technique are described in [33, 34].

With this experimental setup we were able to extract the angle-integrated reaction cross section; the corresponding ^{10}B(p,α_0)^{7}Be S-factor is shown in Figure 1, as black dots. Our data are ≈30% lower than the estimates based on Roughton et al. data [20], even if they agree within the error bars. Consider that in the data of Ref. [20] also the contribution given by the α_1 decay channel is included; anyway, at energies lower than 0.8 MeV, it is of the order of some percent. Vertical bars represent statistical errors; non statistical errors amount to 5.3% and derive from the internal normalization procedure. In Figure 1 we show also results derived from differential cross section taken from the literature. In particular, we used data reported by Brown et al [35] (squares, normalized to our cross section scale), by Cronin [36] and by Overley and Whaling [37]. Considering this comprehensive data set and including also the differential cross section for the ^{10}B(p,p_0)^{10}B elastic scattering [38, 39] and ^{10}B(p,α_1)^{7}Be* reaction [36] data, we performed a simultaneous R-matrix fit of data, with the aim of improving our knowledge of ^{11}C structure. The result of the fit is reported in Figure 1 as solid line. It reproduces quite reasonably the

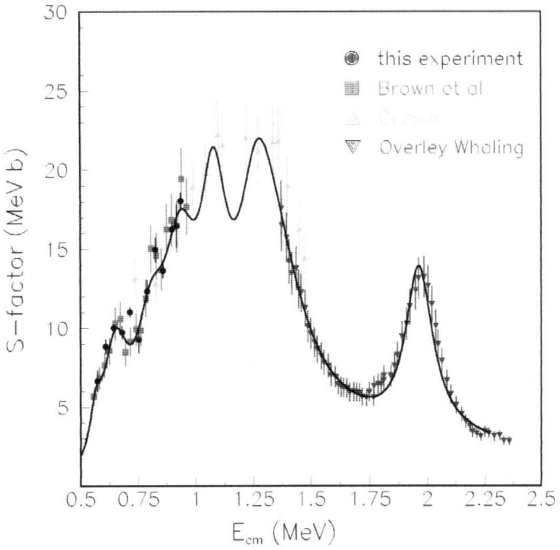

Figure 1. S-factor of the ^{10}B(p,α_0)^{7}Be reaction. Black full dots: data from Ref. [34]. Red squares: S-factor estimate derived from data of Ref. [35], normalized as described in [34]. Green (up) triangles: S-factor estimate derived from data of Ref. [36]. Blue (down) triangles: S-factor estimate derived from data of Ref. [37].

overall data trend only if the presence of a state at 9.36 MeV in ^{11}C level scheme is included, as discussed in details in Ref. [34]. The comparison of angular distributions with R-matrix calculations, that is discussed in details in Ref. [34], suggests a possible $5/2^-$ assignment for this state.

3. The ^{19}F(p,α_0)^{16}O reaction in Nuclear Structure and Astrophysics

The study of low energy ^{19}F(p,α_0)^{16}O reaction allows to investigate the structure of the ^{20}Ne compound nucleus, that can show important α-cluster structures [40]. Anyway, several ambiguities in the ^{20}Ne spectroscopy still persist [22], especially in the lowermost energy region accessible with direct measurements (E_p <0.5 MeV). Furthermore, this reaction is involved in the nuclear astrophysics domain for two reasons. First, it closes the CNOF cycle in the hydrogen-burning phase of massive stars [41]; second, it can be an important fluorine destruction channel in hydrogen-rich stellar environments [27]. Unfortunately, because of the lack of direct low-energy data reported in the literature, the reaction rate estimate at astrophysical energies, based on high-energy extrapolations, is characterized by large uncertainties [18]. Recent indirect data obtained with the Trojan Horse Method pointed out the important role played by a low energy resonance at 113 keV [28, 29]. Considering the importance of this reaction and taking into account the lack of data reported in the literature, we performed two separate experiments at LdA in Napoli ($E_p \approx$0.6-1.0 MeV) and at INFN - Laboratori Nazionali di Legnaro ($E_p \approx$0.2-0.65 MeV) aimed at measuring excitation functions and angular distributions of the ^{19}F(p,α_0)^{16}O reaction in wide (low) energy domain. The details of these experiments and on the devices used for particle detection have been widely discussed in Refs. [24, 26].

The ^{19}F(p,α_0)^{16}O integrated cross section $\sigma(E)$ have been transformed in the astrophysical S-factor, that is shown in Figure 2 as green stars (high energy part of the data set, experiment performed in Napoli) and black triangles (low energy part of the data set, experiment performed in Legnaro). The open triangles, blue diamonds and magenta circles are the result of older direct

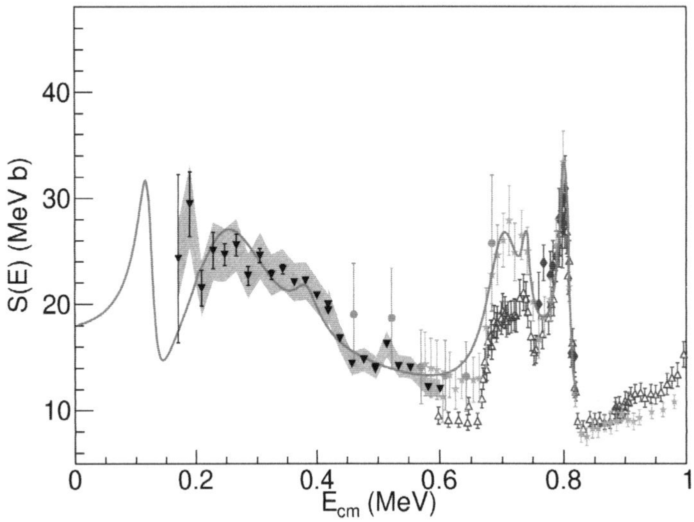

Figure 2. S-factor of the ^{19}F(p,α_0)^{16}O reaction at low energies ($\lesssim$ 1 MeV). Black triangles: experimental data obtained in the Legnaro experiment (vertical bars: statistical errors; azure band: non-statistical errors) [26]. Green stars: data obtained in the Napoli experiment [24]. Light blue triangles: data by Isoya et al [42]. Blue diamonds: data reported by Caracciolo et al [43]. Magenta circles: data by Breuer [44]. Red solid line: R-matrix fit of experimental data obtained in the Legnaro and Napoli experiments [24, 26].

measurements [42, 43, 44], as reported in the NACRE paper [18]. The blue band represents non-statistical errors of the Legnaro experiment, estimated as discussed in Ref. [26]. The results of the two experiments performed in Napoli and Legnaro are in agreement (within the errors) in their overlap region. The inspection of the low energy part of Figure 2 indicates a behaviour of the S-factor that can be attributed to the sum of a direct component and of various resonances of the compound nucleus. An important resonant contribution is due to a broad 2^+ state at 0.251 MeV (E_x=13.095 MeV, Γ=162 keV); this finding is in agreement with considerations made on the angular distributions in Ref. [26]. The red line represents the result of a R-matrix fit on the experimental data. The resonance parameters used are reported in detail in Ref. [26]. We included in this fit procedure the presence of a 2^+ state at 0.113 MeV as suggested by an experiment done with the Trojan Horse Method [28, 29]. The interference effect between the two close-lying 2^+ states explains the asymmetric high energy tail for the 0.251 MeV 2^+ state pointed out by the experimental data.

The low-energy S-factor extrapolated via the R-matrix fit is ≈1.5-2 times larger than the corresponding NACRE evaluation based on the non-resonant extrapolation of high energy data. As a consequence, also the reaction rate is well larger than the NACRE predictions, as discussed in Ref. [26]. This larger reaction rate could lead to a more efficient fluorine destruction in extra-mixing phenomena. This finding is in agreement with recent experimental observations of fluorine abundance in metal-poor AGB stars [27, 45]; therefore, it could contribute to improve the description of fluorine nucleosyntesis in stars [46].

4. Summary

In these proceedings we discuss briefly the results of two experiments concerning the study of the ^{10}B(p,α_0)^{7}Be and ^{19}F(p,α_0)^{16}O reactions at energies near or below the Coulomb barrier.

Their analysis allows to refine the spectroscopy of the formed compound nuclei, i.e. ^{11}C and ^{20}Ne respectively. For the ^{11}C case, a comprehensive R-matrix fit of the ^{10}B(p,α_0)^{7}Be S-factor and of other reaction and scattering channels suggests the presence of an unreported excited state at $E_x \approx 9.36$ MeV, with tentative $J^\pi = 5/2^-$ assignment [34]. Concerning the analysis of the ^{19}F(p,α_0) reaction, we improved the spectroscopy of ^{20}Ne excited states in the $E_x \approx$13-14 MeV domain, as widely discussed in Refs. [24, 26]. Furthermore, we pointed out the role played by low energy resonances (and their interferences) in the S-factor, resulting in a calculated reaction rate at temperatures typical of AGB stars well larger than the commonly adopted extrapolations based on high energy data; this finding seems in the right direction of recent experimental observations of fluorine abundance in metal-poor AGB stars and can be useful to improve our knowledge of fluorine nucleosynthesis in stars.

Acknowledgments
We gratefully acknowledge all the co-authors of Refs. [24, 26, 34] for their valuable help during the data taking and data analysis phases. Thanks are also due to E.Perillo (Napoli) for valuable discussions about the subjects of these proceedings.

References

[1] Cunsolo A *et al.* 1980 *Phys. Rev. C* **21** 2345
[2] Cardella G *et al.* 1987 *Phys. Rev. C* **36** 2403
[3] Rodrigues M R D *et al.* 2014 *J. Phys. Conf Ser.* **569** 012067
[4] Yang Z H *et al.* 2015 *Phys. Rev. C* **91** 024304
[5] Charity R J *et al.* 2008 *Phys. Rev. C* **78** 054307
[6] Dell'Aquila D *et al.* 2016 *Phys. Rev. C* **93** 024611
[7] Dell'Aquila D *et al.* 2016 *EPJ Web of Conf.* **117** 06011
[8] Freer M *et al.* 2011 *Phys. Rev. C* **84** 034317
[9] Avila M L *et al.* 2014 *Phys. Rev. C* **90** 024327
[10] Suzuki D *et al.* 2013 *Phys. Rev. C* **87** 054301
[11] Lombardo I *et al.* 2014 *Nucl. Instrum. Meth. B* **302** 19
[12] Freer M *et al.* 2012 *Phys. Rev. C* **85** 014304
[13] Yamaguchi H *et al.* 2013 *Phys. Rev. C* **87** 034303
[14] Lombardo I *et al.* 2014 *Bull. Russ. Acad. Sc.: Phys.* **78** 1093
[15] Hyldegaard S *et al.* 2010 *Phys. Rev. C* **81** 024303
[16] Burda O *et al.* 2010 *Phys. Rev. C* **82** 015808
[17] Zimmerman W R *et al.* 2013 *Phys. Rev. Lett.* **110** 152502
[18] Angulo C *et al.* 1999 *Nucl. Phys. A* **656** 3
[19] Spitaleri C *et al.* 2014 *Phys. Rev. C* **90** 035801
[20] Roughton N A *et al.* 1979 *At. Data Nucl. Data Tables* **23** 177
[21] De Boer R J *et al.* 2015 *Phys. Rev. C* **91** 045804
[22] Tilley D R *et al.* 1998 *Nucl. Phys. A* **636** 249
[23] Dell'Aquila D and Lombardo I 2016 *J. Phys. Conf. Ser.* **703** 012015
[24] Lombardo I *et al.* 2013 *J. Phys. G* **40** 125102
[25] Lombardo I *et al.* 2014 *J. Phys. Conf Ser.* **569** 012068
[26] Lombardo I *et al.* 2015 *Phys. Lett. B* **748** 178
[27] Lucatello S *et al.* 2011 *Astr. J.* **729** 40
[28] La Cognata M *et al.* 2011 *Astr. J. Lett.* **739** L54
[29] La Cognata M *et al.* 2015 *Astr. J.* **805** 128
[30] Wiescher M *et al.* 1983 *Phys. Rev. C* **28** 1431
[31] Peterson R J *et al.* 1975 *Ann. Phys. Energy* **2** 503
[32] Lombardo I *et al.* 2013 *J. Phys. Conf Ser.* **436** 012021
[33] Lombardo I *et al.* 2016 *EPJ Web of Conf.* **117** 09009
[34] Lombardo I *et al.* 2016 *J. Phys. G* **43** 045109
[35] Brown A B *et al.* 1951 *Phys. Rev.* **82** 159
[36] Cronin J W 1956 *Phys. Rev.* **101** 298
[37] Overley J C and Whaling W 1962 *Phys. Rev.* **128** 315
[38] Chiari M *et al.* 2001 *Nucl. Instrum. Meth. Phys. Res. B* **184** 309

[39] Chiari M *et al.* 2015 *Nucl. Instrum. Meth. Phys. Res. B* **343** 70
[40] Beck C (ed.) *Clusters in Nuclei* Vol. 1,2,3 (Springer)
[41] Rolfs C and Rodney W S 1988 *Cauldrons in the Cosmos* (Chicago Univ. Press)
[42] Isoya A, Ohmura H and Momota T 1958 *Nucl. Phys.* **7** 116
[43] Caracciolo R *et al.* 1974 *Lett. Nuovo Cim.* **11** 33
[44] Breuer G 1959 *Z. Phys.* **154** 339
[45] Abia C *et al.* 2011 *Astr. J. Lett.* **737** L8
[46] Cristallo S *et al.* 2014 *Astr. Astr. A* **46** 570

Phase transitions for rotational states within an algebraic cluster model

E. López Moreno[1], G. E. Morales Hernández[2], P. O. Hess[2] and H. Yépez Martínez[3]

[1] Facultad de Ciencias, Departamento de Física, UNAM. Circuito Exterior, Cd. Universitaria, Circuito Exterior, 04510 México, D.F., Mexico
[2] Instituto de Ciencias Nucleares, UNAM, Circuito Exterior, C.U., A.P. 70-543, 04510 México, D.F., Mexico
[3] Universidad Autónoma de la Ciudad de México, Prolongación San Isidro 151, Col. San Lorenzo Tezonco, Del. Iztapalapa, 09790 México D.F., Mexico

Abstract. The ground state and excited, rotational phase transitions are investigated within the *Semimicroscopic Algebraic Cluster Model* (SACM). The catastrophe theory is used to describe these phase transitions. Short introductions to the SACM and the catastrophe theory are given. We apply the formalism to the case of ^{16}O+$\alpha\rightarrow^{20}$Ne.

1. Introduction

The *Semimicroscopic Algebraic Cluster Model* (SACM) [1, 2] is a successful cluster model for nuclei. It takes into account the Pauli exclusion principle for any cluster model. It was successfully applied to nuclei in the sd shell, where one of the last applications is published in [3]. Cluster systems of astrophysical interest were considered and also spectroscopic factors [4] calculated. A detailed study of quantum phase transitions of the ground state where published in [5, 6]. However, this study was done in a pedestrian way, not using the latest *technology* available, which is the *catastrophe theory* [7]. Excited, rotational states were considered in [8], but also here the treatment was simple.

The objective of this contribution is to apply the *catastrophe theory* to nuclear cluster system, in order to study in a very effective and detailed manner the structure of phase transitions in nuclear cluster systems, in the ground state and for excited states. The *catastrophe theory* was applied with great success, also in nuclear systems [9, 10]. Of special interest, because of its relation to the case presented here, is [11].

2. The Semimicroscopic Algebraic Cluster Model

The basic degrees of freedom are the relative oscillations, described by the $\boldsymbol{\pi}^\dagger_m$ creation operators and by the corresponding annihilation operators $\boldsymbol{\pi}^m$ ($m = \pm 1, 0$). The products of a creation with an annihilation operator form the $U_R(3)$ group, where R refers to the relative motion. A cutoff is introduced by adding a scalar boson, whose creation and annihilation operator are $\sigma^\dagger$ and σ respectively. These bosons do not have a physical meaning but are merely introduced to create a cut-off. With the help of these bosons, one can introduce a cut-off $N = n_\pi + n_\sigma$. The products of all types of boson creation with boson annihilation operators form a $U_R(4)$ group. For a detailed description of the SACM, please consult [1, 2].

The model space is constructed by first calculating the product $(\lambda_1, \mu_1) \otimes (\lambda_2, \mu_2) \otimes (n_\pi, 0)$, where the (λ_k, μ_k) $(k = 1, 2)$ describe the structure of the clusters 1 and 2 and n_π is the relative oscillation number. This results in a sum of many $SU(3)$ irreducible representations (irreps) (λ, μ) with a multiplicity $m_{(\lambda,\mu)}$. This list has to be compared with the content of the shell model, which can be easily obtained by computer routines, which are available to us and can be obtained on request. When a $SU(3)$ irrep in the former mentioned list does not appear in the shell model, this irrep is excluded, while when it appears it is included into the model space of the SACM. In this way, the model space of the SACM is microscopic.

The Hamiltonian, however, is phenomenological and is written as a sum of Casimir operators. When we restrict to spherical clusters, i.e. $(\lambda_k, \mu_k) = (0, 0)$, there is no contributions from the individual clusters. There are two dynamical symmetry group chains of importance, called the $SU_R(3)$ and the $SO_R(4)$ dynamical symmetries. For the case discussed here, the Hamiltonian has the form

$$\mathbf{H} = x\mathbf{H}_{SU(3)} + (1-x)\mathbf{H}_{SO(4)} - \Omega \boldsymbol{H}_{\text{crank}} \tag{1}$$

with

$$\begin{aligned} \boldsymbol{H}_{SU(3)} &= \hbar\omega \boldsymbol{n}_\pi + (a - b\Delta \boldsymbol{n}_\pi)\boldsymbol{C}_2(\boldsymbol{n}_\pi, 0) + \xi \boldsymbol{L}^2 \\ \boldsymbol{H}_{SO(4)} &= \frac{c}{4}[(\boldsymbol{\pi}^\dagger \cdot \boldsymbol{\pi}^\dagger) - (\boldsymbol{\sigma}^\dagger)^2][(\pi \cdot \boldsymbol{\pi}) - (\boldsymbol{\sigma})^2] + \xi \boldsymbol{L}^2 \\ \boldsymbol{H}_{\text{crank}} &= \boldsymbol{L}_{\text{x}} \quad , \end{aligned} \tag{2}$$

where the $\Delta\hat{\mathbf{n}}_\pi = \hat{\mathbf{n}}_\pi - n_0$ is the excitation number of quanta and n_0 is the minimal number of relative oscillation quanta [12], a necessary condition to satisfy the Pauli exclusion principle. The $\boldsymbol{L}_{\text{x}}$ is the x-component of the angular momentum operator and Ω is the frequency parameter which fixes the amount of the angular momentum. The structure of the Hamiltonian is equivalent to introduce a Lagrange multiplier [13].

The semi-classical potential is obtained by first defining a coherent state [14]

$$|\alpha\rangle = \frac{N!}{(N+n_o)!}\mathcal{N}_{N,n_0}\frac{d^{n_0}}{d\gamma^{n_0}}[\boldsymbol{\sigma}^\dagger + (\boldsymbol{\alpha}^* \cdot \boldsymbol{\pi}^\dagger)]^{N+n_0}|0\rangle \quad , \tag{3}$$

where

$$\mathcal{N}_{N,n_0}^{-2} = \frac{(N!)^2}{(N+n_0)!}\frac{d^{n_0}}{d\gamma_1^{n_0}}\frac{d^{n_0}}{d\gamma_2^{n_0}}[1 + \gamma_1\gamma_2(\alpha^* \cdot \alpha)]^{N+n_0} \tag{4}$$

is the normalization constant. The α_m $(m = \pm 1, 0)$ are the coherent state parameters. The γ_k are set to zero after having applied the derivatives.

Finally, calculating the expectation value of the total Hamiltonian, i.e., $\langle \alpha \mid \boldsymbol{H} \mid \alpha \rangle$, the semi-classical potential is obtained. When $\Omega = 0$, then the only relevant parameter is $\alpha = \alpha_0$, describing the relative distance of the two clusters [14, 5, 6]. However, in general we need all three components when cranking is included. For reason of simplicity, we use polar coordinates, i.e.

$$\begin{aligned} \alpha_{\pm 1} &= \tfrac{\alpha}{\sqrt{2}} e^{\pm i\phi} \sin(\theta) \\ \alpha_0 &= \alpha \cos(\theta) \quad . \end{aligned} \tag{5}$$

The potential is obtained through the geometrical mapping. We add to the geometrical potential constant terms such that for $\alpha \to 0$ the potential approaches zero. The final rresult is:

$$\begin{gathered}\widetilde{V} = A\left(\alpha^2 \frac{F_{11}(\alpha^2)}{F_{00}(\alpha^2)} - \frac{n_0}{N+n_0}\right) + \\ (B + C\sin^2(2\theta))\left(\alpha^4 \frac{F_{22}(\alpha^2)}{F_{00}(\alpha^2)} - \frac{n_0(n_0-1)}{(N+n_0)(N+n_0-1)}\right) + \\ D\cos(2\theta)\,\alpha^2\,\frac{F_{20}^{N-2}(\alpha^2)}{F_{00}(\alpha^2)} + \\ \left(\alpha^6 \frac{F_{33}(\alpha^2)}{F_{00}(\alpha^2)} - \frac{n_0(n_0-1)(n_0-2)}{(N+n_0)(N+n_0-1)(N+n_0-2)}\right) - \\ \Omega(\Omega')\sin(2\theta)\cos(\phi)\left(\alpha^2 \frac{F_{11}(\alpha^2)}{F_{00}(\alpha^2)} - \frac{n_0}{N+n_0}\right). \end{gathered} \tag{6}$$

The *Control Parameters*, $c_i = \{A, B, C, D, \Omega\}$, of the SE are functions of the parameters in the Hamiltonian.

As these control parameters varies, a complete family of analytic functions, $\widetilde{V}(\alpha, \theta, \phi : c_i)$, is generated.

$$A(x) = \\ \frac{4x(\hbar\omega/4 + a - b + bn_0) + 2\xi - (1-x)\frac{c}{2}(N + n_0 - 1)}{(N + n_0 - 1)(N + n_0 - 2)(-x\,b)},$$

$$\begin{aligned} B(x) &= \frac{x\,(a - b\,(6 - n_0)) + (1-x)\,\frac{c}{2}}{(N + n_0 - 2)(-xb)}, \\ C(x) &= \frac{\xi - (1-x)\,\frac{c}{4}}{(N + n_0 - 2)(-x\,b)}, \\ D(x) &= \frac{(1-x)\,\frac{c}{2}}{(N + n_0 - 2)(-x\,b)}, \\ \Omega(\Omega', x) &= \frac{\Omega'}{(N + n_0 - 1)(N + n_0 - 2)(-x\,b)}. \end{aligned} \tag{7}$$

This potential is the subject of investigation in this contribution. In the next sections the structure under phase transitions is investigated.

3. Catastrophe Theory and phase transitions

The semiclassical behavior of the system can be studied by the methods of catastrophe theory. It is useful to consider only the value $\phi = 0\,(or\ \pi)$, and the change of variables, $v = \cos 2\theta$. Then the potential, Eq.(6), can be written as

$$\widetilde{V} = \frac{\alpha^2}{q_0(\alpha)}\,W_0(\alpha, v; c_i). \tag{8}$$

The $c_i = \{A, B, C, D, \Omega\}$, are the control parameters, $q_0(\alpha)$ is a polynomial in even powers, positive integers, $q_0(0) \neq 0$, and

$$\begin{aligned} W_0 = &\left(A \mp \Omega\sqrt{1 - v^2}\right) p_A(\alpha) + D\,v\,p_D(\alpha) \\ &(B + C\,(1 - v^2))\;p_B(\alpha) + p_0(\alpha). \end{aligned} \tag{9}$$

Journal of Physics: Conference Series **730** (2016) 012017 doi:10.1088/1742-6596/730/1/012017

Here, for brevity, we omit the arguments into the adjustable control parameters in Eq.(7), and the variables α, and v, are derived from those in Eq.(5). For any values of N and n_0 chosen, the $\{p_X(\alpha)\}$ in Eq.(9), results in a set of polynomials in even powers of α, positive integer coefficients, and a non-zero constant, *i. e.*, $p_X(0) \neq 0$.

In order to study the stability properties of the semiclassical system, we first evaluate the critical points of potential, $\widetilde{V}$, by means of its first derivatives: $U_\alpha \equiv \frac{\partial \widetilde{V}}{\partial \alpha}$, and $U_v \equiv \frac{\partial \widetilde{V}}{\partial v}$. We find that these can be factored as:

$$U_\alpha = \frac{\alpha}{q_1(\alpha)} W_1(\alpha, v; c_i)\,, \tag{10}$$

where, $q_1(\alpha) = \kappa_1 \left(\frac{q_0(\alpha)}{(1+\alpha^2)}\right)^2$, κ_1 a constant, and

$$U_v = \frac{\alpha^2}{q_0(\alpha)} \frac{\partial W_0}{\partial v}\,. \tag{11}$$

We can see from these last expressions that when the variable α, takes the value, $\alpha_0 = 0$, it is a fundamental root; that is, it is a critical point for the potential $\widetilde{V}$, for any choice of the control parameters. A degeneracy in any critical point means the coalescence of two or more of them due to a variation of control parameters. In this situation, the stability of the system can not be studied by means of the usual second derivatives criteria. The shapes and stability of the semiclassical system can be studied by means of the catastrophe theory program, as considered in the past [10], as follows:

(i) The fundamental root, $\alpha_0 = 0$, is selected in order to obtain the *germ* of the potential $\widetilde{V}$, that is, to find the point within the control parameter space where the maximum degeneracy is presented. This *germ* is the first term of the Taylor series expansion of $\widetilde{V}$ which can not be cancelled by any set of control parameters. The first terms, in increasing order in the Taylor expansion, are succesively eliminated by means of a set of relations among the control parameters. This process permits us the determination of the essential parameters.

(ii) One constructs the Separatrix of the system. The first part of this separatrix consist of the bifurcation sets of the energy surface. These are the *loci* within the essential parameter space where a transition occurs from one minimum to another. The bifurcation set satisfies the condition that the matrix of second derivatives of the energy surface, when evaluated at the critical points, has a null determinant value. However, in this work, in order to evaluate the bifurcation set we shall employ an important topologic feature of catastrophe theory instead [7]: the mapping, from the critical manifold into the essential parameter space, becomes singular at the bifurcation set. The critical manifold is the surface $\{(x_{cr}^{(j)}, c_i)\}$, of those points obtained when all the c_i's are continuously varied, for every j-critical point, $(x_{cr}^{(j)}, c_i)$. The mapping mentioned becomes singular if the Jacobian of the transformation is zero, and in general the mapping is invertible except for the set points at which the tangent plane to the manifold is vertical, meaning that they have associated the same control parameters, that is when the critical points present a coalescence.

(iii) The Maxwell sets are determined. These sets constitute the locus of points in the essential parameter space for which the energy surface takes the same value in two or more critical points. They can be found through the Clausious-Clapeyron equations [10]; however, due to the big magnitude of the problem in the present case, we shall present an original approach when solving this problem below.

(iv) Finally, the separtrix shall be constructed by the union of the bifurcation and the maxwell sets. This procedure divides the essential parameter space into shape stability regions and identifies the loci where there are transitions according to its order.

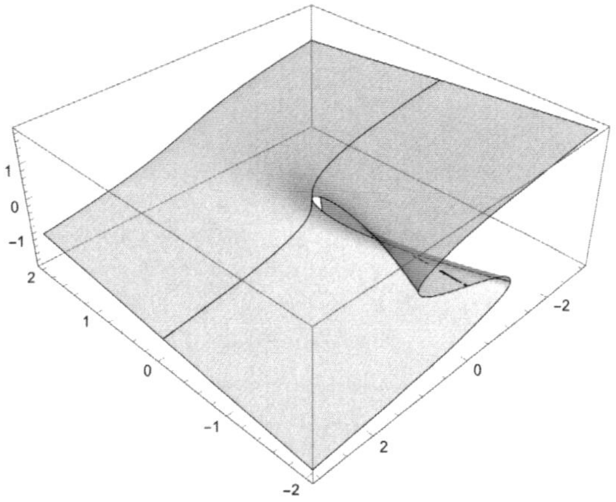

Figure 1. The Cusp catastrophe. For the cusp potential $V_{cusp}(x) = x^4/4 + ax^3 + bx + c$, the set of all its critical points x_{cr}: $dV_{cusp}(x; a, b)/dx = x^3 + ax + b = 0$, when the essential parameters (a, b) change, spans the critical manifold of the Cusp: (a, b, x_{cr}).

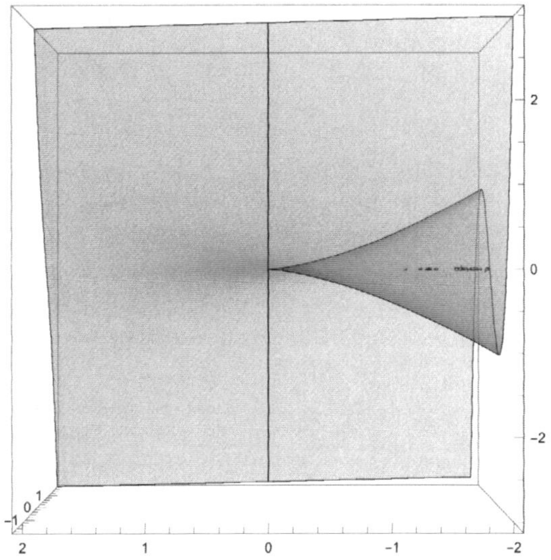

Figure 2. Mapping the Cusp. A Cusp is a mapping singularity when the manifold in Fig.(1) is projected onto the essential parameters space. The Separatrix: $(\frac{a}{3})^3 + (\frac{b}{2})^2 = 0$.

4. Application to $^{16}O + \alpha \rightarrow {}^{20}Ne$

Next we follow the procedure indicated above, and evaluate the numerical results for the case of ^{16}O+$\alpha \rightarrow ^{20}$Ne. We shall include, $N = 12$, and $n_0 = 8$, values into the previous expressions in what follows.

First, we evaluate the Taylor series expansion of the energy surface Eq.(8), around the fundamental root, $\alpha_0 = 0$ ($\phi = 0$):

$$\widetilde{V} = T_2\,\alpha^2 + T_3\,v\,\alpha^2 + T_4\,\alpha^4 + T_5\,v\,\alpha^4 + T_6\,\alpha^6 + \cdots, \tag{12}$$

Journal of Physics: Conference Series **730** (2016) 012017 doi:10.1088/1742-6596/730/1/012017

with the succesive coefficients:

$$\begin{aligned} T_2 =& \; 18\,(28 + 57\,\rho_2 + 48\,\rho_1)/95\,, \\ T_3 =& \; 594\,D/19\,, \\ T_4 =& \; 18\,(28 + 57\,\rho_2 + 48\,\rho_1)/95\,, \\ T_5 =& \; -508464\,D/19\,, \\ T_6 =& \; 144\,(809893 + 1930077\,\rho_2 + 1519398\,\rho_1)/19\,, \\ \rho_2 \equiv& \; (A - \Omega) \\ \rho_1 \equiv& \; (B + C)\,. \end{aligned} \tag{13}$$

The first order term in Eq.(13) is obviously eliminated because it is evaluated at a critical point.

The T_2 term is eliminated at the points of the following hyper surface in the parameter space

$$\mathcal{C}_0 = \left\{ (\rho_2, \rho_{1c}) \,\middle|\, \rho_{1c}(\rho_2) = -\frac{19}{16}\rho_2 - \frac{7}{12} \right\}. \tag{14}$$

At $\mathcal{C}_0$ the fundamental root becomes double degenerate because their second derivatives are cancelled.

The T_3 term cancels, and thus the fundamental root will degenerate further, for those points in parameter space laying on the plane $\mathcal{D}_c$, defined by:

$$D = 0\,. \tag{15}$$

On this plane, variable-v becomes double-degenerated, as we shall demonstrate later. It is important to observe that the plane $\mathcal{D}_c$ corresponds to the limit $x = 1$, as it is clear from Eq.(7), that is to the limit $SU(3)$ in Eq.(1).

The T_4 term, corresponding to the fourth order derivatives evaluated at the fundamental root, is eliminated if Eqs.(14) and (15) are satisfied, and simultaneously the following point is chosen:

$$(\rho_2, \rho_{1c}(12/19)) = (12/19, -4/3)\,, \tag{16}$$

increasing to the fourth order the α degeneracy. Under these conditions, the restriction

$$\Omega \to \Omega_c \equiv \frac{32}{19}\,C\,, \tag{17}$$

shall further increase the degeneracy of the v-variable to fourth order, being $v_0 = 0$ the unique variable-v critical point, as we shall show later.

The term T_5 will be then automatically cancelled.

It is important to note that when conditions, Eq.(14) to Eq.(17), are satisfied, then it is imposible to cancel the sixth order term in Taylor series expansion by any further parameters relation. In this way, we have obtained the *germ* of the energy potential as that point $\mathcal{G}_0$: $(\rho_{2G}, \rho_{1G}) = (12/19, -4/3)$ within the parameter space satisfying all the above conditions. When evaluated at this *germ* in the parameter space, the potential becomes: a) six-fold degenerated at its minimum, b) parameter independent, and c) a ratio of two polynomials, $\widetilde{V}(\alpha, v = 0) = \mathcal{P}_n(\alpha)/\mathcal{P}_d(\alpha)$; with, $\mathcal{P}_n(0) = 0$, $\mathcal{P}_d(0) \neq 0$, and $\widetilde{V}(\alpha \to \infty, v) = 11/57$ (see Fig.(3)). At its *germ* point, the potential $\widetilde{V}$ has no other minimal critical point, and its maximum is attained only at infinity. If any of the parameters vary, going out of this *germ*, then other shapes and phase-shapes-transitions can occur.

From the above discussion we found the *germ*, $\mathcal{G}_0$, and conclude that the separatrix for the fundamental root consist of the plane $\mathcal{D}_c$, Eq(15), and the straight line $\mathcal{C}_0$, Eq(14). On the one hand, out of separatrix $\mathcal{D}_c$, as it will be shown later, the doubly degenerate $v_0 = 0$ critical point

Journal of Physics: Conference Series **730** (2016) 012017 doi:10.1088/1742-6596/730/1/012017

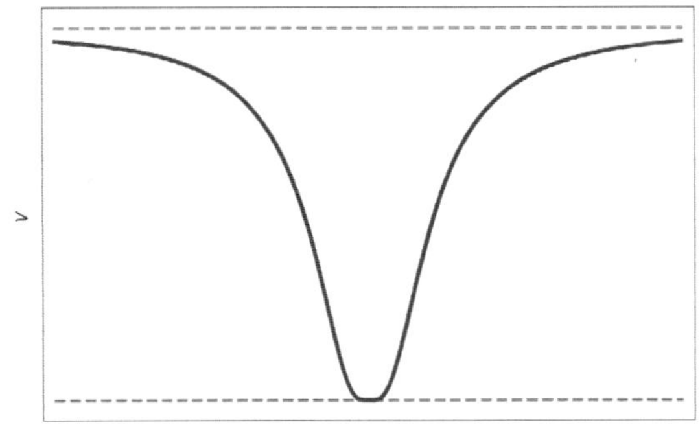

Figure 3. Energy Surface $\widetilde{V}(\alpha, v)$, evaluated at the critical *germ* in the essential parameter space. The minimum at the origin cannot be approximated by an harmonic oscillator. It has an asymptotic value, $\widetilde{V}(\alpha \to \infty, v = 0)$, indicated as the upper horizontal line.

bifurcates, and $v = 0$ is no longer an extreme value; this means a shape transition in variable v-direction for the fundamental root. On the other hand, at the separatrix line $\mathcal{C}_0$ within the plane $\mathcal{D}_c$ there is a shape transition in the variable α. Crossing this line in one direction (from above) a minimum changes to a maximum. If one goes through a continuous change within the line $\mathcal{C}_0$, then there is a shape transition in the variable α. Crossing the *germ* in one direction (from left): a minimum changes to a maximum.

In this contribution we shall be interested in points very near to the separatrix plane $\mathcal{D}_c$, so we consider parameter values $D \sim 0$; however, we shall study the more general bifurcation and the Maxwell sets in the variable α, for all α-extreme values with $v_{cr} = 0$. At the last part of this work, a very brief numerical example of bifurcation in the v-critical point shows us that a spontaneous symmetry breaking phenomena is represented as a phase transition, and also we present a numerical example of the existence of Maxwell points not including the fundamental root.

Bifurcation set within $\mathcal{D}_c$ plane.- In this section we take $v_0 = 0$ critical point only. Then, $W_0 \to w_0(\alpha; \rho_2, \rho_1)$, and we can write Eqs.(9), and (10) as:

$$w_0(\alpha; \rho_2, \rho_1) = \rho_2\, p_A(\alpha) + \rho_1\, p_B(\alpha) + p_0(\alpha), \tag{18}$$

$$U_\alpha = \frac{\alpha}{q_1(\alpha)}\, w_1(\alpha; \rho_2, \rho_1). \tag{19}$$

We want to evaluate the bifurcation set of the $\alpha_c \neq 0$ critical points. These critical points are the roots of the equation

$$w_1(\alpha; \rho_2, \rho_1)\Big|_{\alpha_c} = 0. \tag{20}$$

As we mention in Sect. III.(ii), we begin by considering the mapping, $\mathcal{M}$, from the critical manifold to the essential parameter space

$$\mathcal{M}: \qquad \{(\alpha_c, \rho_2, \rho_1)\} \longrightarrow \{(\rho_2, \rho_1)\}. \tag{21}$$

In order to find the singular points at the parameter space of this mapping, instead of solving for α_c in Eq.(20), we solve for ρ_1, and rename as follows:

$$\alpha_c \to \lambda_1, \quad \rho_2 \to \lambda_2, \quad \rho_1 \to S_1(\lambda_1, \lambda_2). \tag{22}$$

Journal of Physics: Conference Series **730** (2016) 012017 doi:10.1088/1742-6596/730/1/012017

The mapping is singular if its Jacobian determinant is zero, or equivalently, if

$$\frac{\partial S_1(\lambda_1,\lambda_2)}{\partial \lambda_1} = 0\,. \tag{23}$$

Finally, we solve for λ_2 in this last equation; this solution gives a critical λ_2 as a function of λ_1:

$$\lambda_{2c} \equiv \mathcal{S}_2(\lambda_1)\,. \tag{24}$$

The mapping $\mathcal{M}$ in Eq.(21) is invertible, except for a smooth curve on the critical manifold in the tridimensional space, given by:

$$\{(\alpha_c,\rho_2,\rho_1)\} = (\lambda, \mathcal{S}_2(\lambda), \mathcal{S}_1(\lambda)) \tag{25}$$

where we have defined

$$\mathcal{S}_1(\lambda) \equiv S_1(\lambda, \mathcal{S}_2(\lambda))\,. \tag{26}$$

The separatrix, $\mathcal{C}_1$, for the critical points $\alpha_c \neq 0$, is the set of points on the essential parameter space obtained parametrically as:

$$\mathcal{C}_1 : \qquad \left\{ \left(\mathcal{S}_2(\lambda), \mathcal{S}_1(\lambda)\right) \Big|\, \lambda \,\epsilon\, \mathcal{R}e^+ \right\}. \tag{27}$$

This separatrix, $\mathcal{C}_1$, curve surges from the *germ* at the cusp point:

$$\left(\mathcal{S}_2(0), \mathcal{S}_1(0)\right) = (12/19, -4/3)\,, \tag{28}$$

as a numerical evaluation can demonstrate.
Maxwell set $\mathcal{C}_M$ within $\mathcal{D}_c$ plane.- In the first place, we observe that for a given point on the essential parameter space, we obtain a particular graph of potential $\widetilde{V}$. Further, suppose that we have chosen a point belonging to the Maxwell set. Then, at this particular point (ρ_2, ρ_1), the form of the graph is such that both, the fundamental critical point $\alpha_0 = 0$, and another $\alpha_c \neq 0$ critical point, satisfy the equality

$$\widetilde{V}(\alpha_0;\rho_2,\rho_1) = \widetilde{V}(\alpha_c;\rho_2,\rho_1) = 0\,. \tag{29}$$

Also, it is clear from Eqs.(8), and (18), that the following equation is satisfied

$$w_0(\alpha;\rho_2,\rho_1)\Big|_{\alpha_c} = 0\,. \tag{30}$$

At the second place, we observe that in a more general (ρ_2, ρ_1) point, a root $\alpha^{(j)} \neq 0$ of the equation:

$$w_0(\alpha;\rho_2,\rho_1) = 0\,. \tag{31}$$

means that the graph of potential $\widetilde{V}$ is crossing the horizontal α-axis, at a distance $\alpha^{(j)}$-away from the origin. As the (ρ_2, ρ_1) point changes towards a Maxwell point, two different roots, $\alpha^{(j)}$ and $\alpha^{(k)}$, of this last equation coalesce. Obviously, this also means that at the coalescence $\alpha^{(j)}$-value the α-horizontal axis becomes a tangent, and thus a critical Maxwell point will be found. From the facts pointed in this paragraph, we propose a method, inspired in catastrophe theory, that produces the Maxwell set as the singularities on the parameter space (ρ_2, ρ_1) of a mapping from the three-dimensional surface, $(\alpha^{(j)};\rho_2,\rho_1)$, of all different $\alpha^{(j)}$-root points of Eq.(31), obtained by a continuous variation of the essential parameters, similar to the singularity illustrated in Fig.(2).

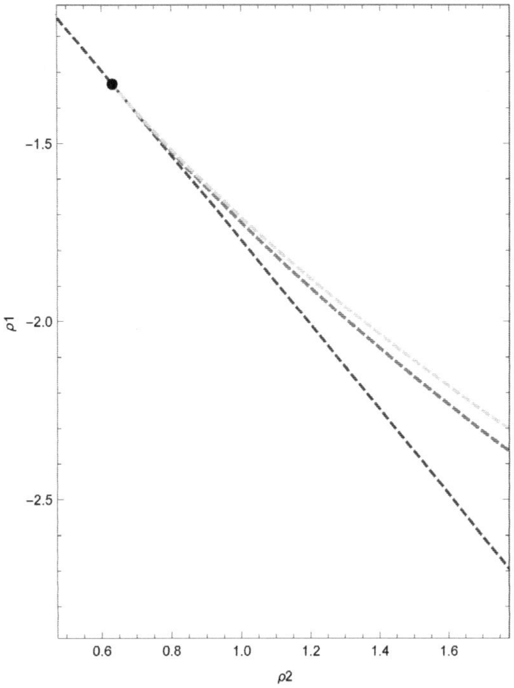

Figure 4. Separatrix. The straight line $\mathcal{C}_0$ is a bifurcation set for the fundamental root, $\alpha_0 = 0$. The upper line, $\mathcal{C}_1$, is a bifurcation set for critical points, $\alpha_c \neq 0$. The intermediate line, $\mathcal{C}_M$ is a Maxwell set. The cusp, signaled a big dot at the point (12/19, -4/3), is the locus of the *germ* $\mathcal{G}_0$. This *germ* is in itself a separatrix within the subspace $\mathcal{C}_0$: it divides the behavior of the fundamental root, being a minima at any point from above and, going through a phase transition at this *germ*, it turnes to maxima at the other side.

The coalescence points of the roots $\alpha^{(j)} \neq 0$ of Eq.(31) can be found by considering the singularities of the mapping:

$$\mathcal{M}_M : \{(\mu_1, \mu_2, T_1(\mu_1, \mu_2))\} \longrightarrow \{(\rho_2, \rho_1)\} . \tag{32}$$

$T_1(\mu_1, \mu_2)$ obtained solving for ρ_1 in Eq.(31), and being renamed as follows:

$$\alpha^{(j)} \to \mu_1 , \quad \rho_2 \to \mu_2 , \quad \rho_1 \to T_1(\mu_1, \mu_2) . \tag{33}$$

The mapping is singular if its Jacobian determinant is zero, or equivalently, if

$$\frac{\partial T_1(\mu_1, \mu_2)}{\partial \mu_1} = 0 . \tag{34}$$

We obtain the critical μ_2, by solving it from this last equation as a function of μ_1:

$$\mu_{2c} \equiv \mathcal{T}_2(\mu_1) . \tag{35}$$

The separatrix, $\mathcal{C}_M$, for the roots $\alpha^{(j)} \neq 0$, is the loci on the essential parameter space obtained as:

$$\mathcal{C}_M : \left\{ \Big(\mathcal{T}_2(\mu), \mathcal{T}_1(\mu)\Big) \Big| \, \mu \,\epsilon\, \mathcal{R}e^+ \right\} ; \tag{36}$$

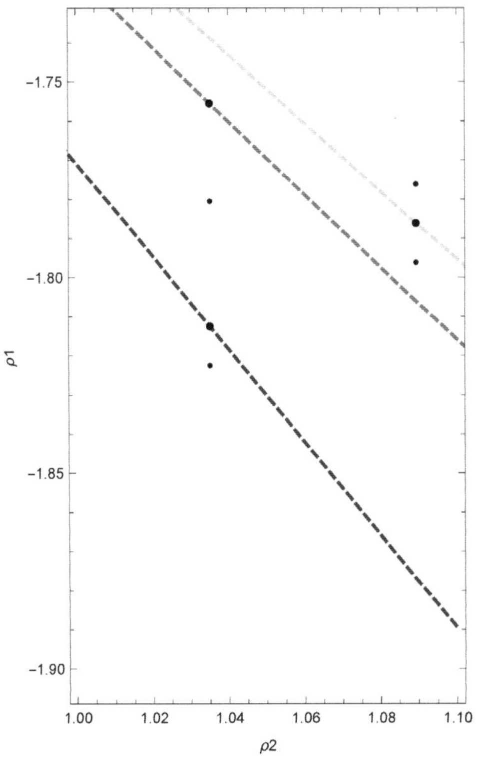

Figure 5. Separatrix. Selected points at the separatrix within the subspace $\mathcal{D}_c$: it divides the behavior of the fundamental root, and other $\alpha_c \neq 0$ critical points, as it is shown in the next figures.

where we have defined

$$\mathcal{T}_1(\mu) \equiv \mathcal{T}_1(\mu, \mathcal{T}_2(\mu))\,. \tag{37}$$

This Maxwell curve surges from the *germ* at the cusp point:

$$\Big(\mathcal{T}_2(0), \mathcal{T}_1(0)\Big) = (12/19, -4/3)\,, \tag{38}$$

as a numerical evaluation can demonstrate (see Fig. 4).

The parametric plotting on the essential parameter space of the bifurcation sets: $\mathcal{C}_0$ (Eq. (14)), $\mathcal{C}_1$ (Eq. (27)), and the Maxwell set $\mathcal{C}_M$ (Eq. (36)), defines the Separatrix of the system. The character of each equilibrium points remain the same at each distinct region, and a sudden qualitative changes take place when this separatrix is crossed as a result of a general variation of the Hamiltonian parameters. These sudden changes are treated as a mathematical catastrophes, and constitute the quantum phase transitions of the ground and the excited states for the cluster model Hamiltonian in Eq. (1).

In Fig.(5) we choose a point at every region of the Separatrix, and at every component of this Separatrix. In the following figures (Fig.(6) to Fig.(8)), we plot the corresponding potentials.

5. Conclusions

In this work, by means of coherent states and the catastrophe theory, the separatrix of an algebraic cluster model was constructed. We have discussed the salient features that allow to discriminate between a region in which phase transitions appear and a region that exhibits

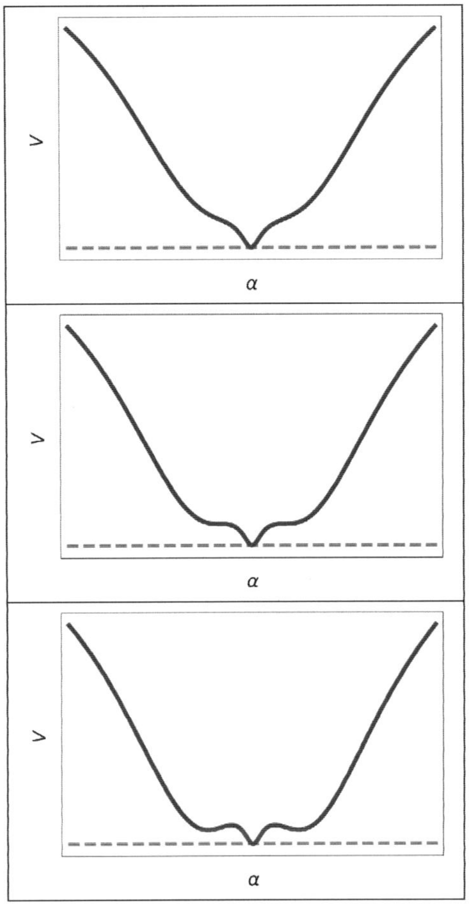

Figure 6. Shape transition at bifurcation set $\mathcal{C}_1$, at the parameter points indicated in Fig.5. The top figure shows a minimun at the $\alpha = 0$ coordinates origin. This is the general behavior at the upper region, above the $\mathcal{C}_1$ set. The middle figure corresponds to a point on this set; there, a critical degenerated saddle point appears. The bottom figure shows the behavior of the potential at the region in between the $\mathcal{C}_1$, and the $\mathcal{C}_M$ sets: a metastable $\alpha \neq 0$ appears.

coexisting shapes of this model. We established the germ point on the essential parameter space. When evaluated at this *germ* in the parameter space, the potential becomes: a) six-fold degenerated at its minimum, b) parameter independent, and c) a ratio of two polynomials. At its *germ* point, the potential $\widetilde{V}$ has no other minimum critical point, and its maximum is attained only at infinity. If any of the parameters vary, going out of this *germ*, then other shapes and phase-shapes-transitions can occur. The *germ* is the locus of the separatrix cusp point, and by crossing this point we observe a sixth order phase transition, the highest one in this model. Here we limit our attention to regions very near the $SU(3)$ limit on the model, and then the fundamental root of the potential play a central role. In a future work, it remains to consider equilibria minimal points far from the origin, in order to extend the Maxwell set of shape coexistence forms.

Acknowledgments
P. O. H. acknowledges financial support from DGAPA (PAPIIT IN 100315).

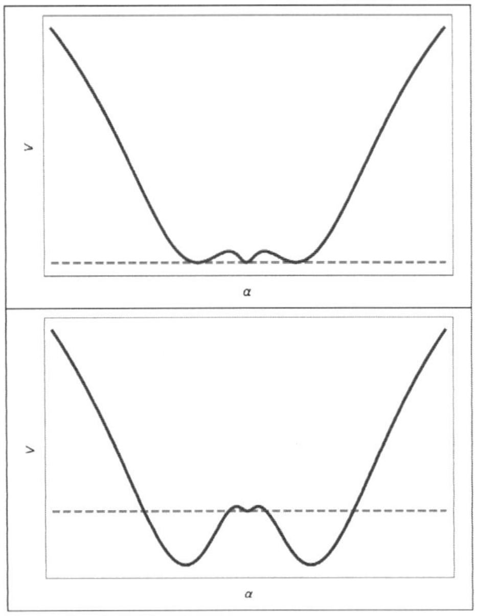

Figure 7. Shape transition at Maxwell set $\mathcal{C}_M$, at the parameter points indicated in Fig.5. The top figure exhibits the coexistence of two minima at the Maxwell set. At the bottom figure, the potential is evaluated at the point between the $\mathcal{C}_M$ and the $\mathcal{C}_0$ sets. There, the fundamental root corresponds to an excited metastable state, and the ground state is an $\alpha \neq 0$ deformed state.

References

[1] Cseh J 1992 *Phys. Lett. B* **281** 173
[2] Cseh J and Lévai G 1994 *Ann. Phys. (N.Y.)* **230** 165
[3] Yépez-Martínez H, Fraser P R, Hess P O and Lévai G 2012 *Phys. Rev. C* **86** 034309
[4] Hess P O, Algora A, Cseh J and Draayer J P 2004 *Phys. Rev. C* **70** 051303(R)
[5] Yépez-Martínez H, Fraser P R, Hess P O and Lévai G 2012 *Phys. Rev. C* **85** 014316
[6] Fraser P R, Yépez-Martínez H, Hess P O and Lévai G 2012 *Phys. Rev. C* **85** 014317
[7] Gilmore R 1981 *Catastrophe Theory for Scientists and Engineers* (New York, Wiley)
[8] Morales-Hernández G E, Yépez-Martínez H and Hess P O 2012 *J. Phys.: Conf. Ser.* **387** 012019
[9] López-Moreno E 1998 *Deformaciones nucleares y estabilidad de modelos colectivos* (Ph.D. Thesis, Universidad Nacional Autónoma de México)
[10] López-Moreno E and Castaños O 1996 *Phys. Rev. C* **54** 2374
[11] López-Moreno E, Grether M and Velázquez V 2011 *J. Phys. A* **44** 475301
[12] Wildermuth K and Tang Y C 1977 *A Unified Theory of the Nucleus* (Braunschweig, Friedr. Vieweg & Sohn Verlagsgesselschaft mbH)
[13] Ring P and Schuck P 1980 *The Nuclear Many-Body Problem* (Heidelberg, Springer)
[14] Hess P O, Lévai G and Cseh J 1996 *Phys. Rev. C* **54** 2345

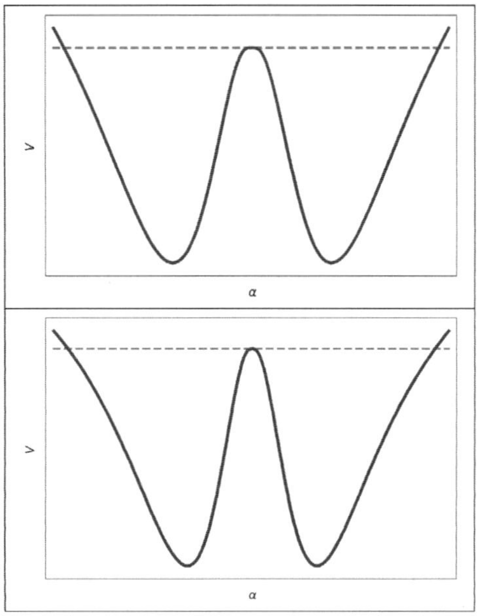

Figure 8. Shape transition at the $\mathcal{C}_0$ set in Fig.5. The top figure exhibits degenerated maximum at the origin and a deformed minimum at $\alpha_c \neq 0$. The bottom figure shows the behavior for all those points bellow the $\mathcal{C}_0$ set: The potential has only one $\alpha_c \neq 0$ minimum and a regular maximum at the origin.

Journal of Physics: Conference Series **730** (2016) 012018 doi:10.1088/1742-6596/730/1/012018

On the robustness of the r-process in neutron-star mergers against variations of nuclear masses

J J Mendoza-Temis[1,2], M R Wu[3], G Martínez-Pinedo[3,4], K Langanke[3,4], A Bauswein[5,6], H-T Janka[7] and A Frank[1,2]

[1]Instituto de Ciencias Nucleares, Universidad Nacional Autónoma de México, 04510 México, D.F., México
[2]Centro de Ciencias de la Complejidad, Universidad Nacional Autónoma de México, 04510 México, D.F., México
[3]Institut für Kernphysik (Theoriezentrum), Technische Universität Darmstadt, Schlossgartenstraße 2, 64289 Darmstadt, Germany
[4]GSI Helmholtzzentrum für Schwerionenforschung, Planckstraße 1, 64291 Darmstadt, Germany
[6]Heidelberger Institut für Theoretische Studien, Schloss-Wolfsbrunnenweg 35, 69118 Heidelberg, Germany
[5]Department of Physics, Aristotle University of Thessaloniki, 54124 Thessaloniki, Greece
[7]Max-Planck-Institut für Astrophysik, Postfach 1317, 85741 Garching, Germany

E-mail: joel.mendoza@correo.nucleares.unam.mx

Abstract. r-process calculations have been performed for matter ejected dynamically in neutron star mergers (NSM), such calculations are based on a complete set of trajectories from a three-dimensional relativistic smoothed particle hydrodynamic (SPH) simulation. Our calculations consider an extended nuclear reaction network, including spontaneous, β- and neutron-induced fission and adopting fission yield distributions from the ABLA code. In this contribution we have studied the sensitivity of the r-process abundances to nuclear masses by using diferent mass models for the calculation of neutron capture cross sections via the statistical model. Most of the trajectories, corresponding to 90% of the ejected mass, follow a relatively slow expansion allowing for all neutrons to be captured. The resulting abundances are very similar to each other and reproduce the general features of the observed r-process abundance (the second and third peaks, the rare-earth peak and the lead peak) for all mass models as they are mainly determined by the fission yields. We find distinct differences in the predictions of the mass models at and just above the third peak, which can be traced back to different predictions of neutron separation energies for r-process nuclei around neutron number $N = 130$.

1. Introduction

The astrophysical r process produces about half of the heavy elements in the Universe, including all of the actinides [1, 2]. It is commonly accepted that it occurs as a sequence of neutron captures and β decays in environments with extreme neutron densities. The natural candidate involves NSM, in fact numerical simulations indicate that the matter ejected during the dynamical phase is very neutron rich with extremely large neutron-to-seed ratios ($R_{n/s} > 400$) [3, 4, 5]; i.e. there are many neutrons which can be captured by seed nuclei transporting matter to very heavy nuclei in the region of the nuclear chart where decay by fission is possible. In this study, we work under the assumption that all ejecta remain neutron rich.

Nuclear masses are particularly important as they, via the neutron separation energies, define the r-process path in the nuclear chart and secondly they are crucial ingredients in the statistical model calculation of neutron capture cross sections. As the masses of the extremely neutron-rich nuclei on the r-process path are not known experimentally, they have to be modeled. We use four different mass models (FRDM [6], HFB21 [7], WS3 [8] and DZ31 [9, 10]) for the calculation of the neutron capture rates that enter in the r-process simulations. Despite there are two recent sensitivity studies already available, on the one hand randomly varying individual masses [11], and on the other hand exploring systematic uncertainties for masses estimated via different energy density functionals [12]; in this contribution we aim at determining the nuclear origin of the robust r-process pattern observed in several NSM simulations. It is often stated that NSM produce a robust r-process due to fission cycling. However, this statement simply expresses the fact that fission cycling is unavoidable due to the large neutron-to-seed ratios ($R_{n/s}$) reached in NSM ejecta without really explaining the nuclear mechanism responsible for the robustness. Our paper is organized as follows. In the next section we give a brief description of our r-process simulations and the input being used. The results of our simulations for the r-process abundances and their dependence on the adopted mass models are presented and discussed in section 3. Finally, we conclude in section 4.

2. NSM trajectories and nuclear input

We have calculated the r-process abundances for 528 trajectories with a total ejected mass of $\sim 1.70 \times 10^{-3}$ M$_\odot$ comming from a binary system of two NSs with gravitational masses of 1.35 M$_\odot$, which may be representative for the observed double NS systems (see e.g. [13] for a compilation of measured binary NS masses). In a previous work [14] such trajectories were classified with respect to two competing rates: the depletion rate due to neutron captures on seed nuclei (λ_n) and the hydrodynamic expansion rate (λ_d). "Slow ejecta" corresponds to most of the trajectories (484 trajectories with a total mass of $\sim 1.57 \times 10^{-3}$ M$_\odot$), where nearly all initial neutrons can be captured as $\lambda_n \gtrsim \lambda_d$ until the end of the r process at ~ 1 s. On the other hand, about $\sim 10\%$ of the ejecta (44 trajectories, $\sim 1.28 \times 10^{-4}$ M$_\odot$) initially expand extremely fast ("fast ejecta"), and as a consequence free neutrons are left at the end of the r process. We have started our r-process calculations at temperatures of $T = 6$ GK, with densities ranging from $\rho \sim 10^7$ g cm^{-3} to $\sim 3 \times 10^{13}$ g cm^{-3}. The initial matter compositions have been determined assuming the matter to be in Nuclear Statistical Equilibrium (NSE). The initial neutron-to-seed ratios ($R_{n/s}$) of those ejecta range from 400 to 2000. Starting from these initial compositions we have followed the r-process evolution by a large network including more than 7300 nuclei which cover the nuclear chart from free nucleons up to ^{313}Ds. The dynamics of the r-process was governed by the astrophysical trajectories, however, consistently corrected for reheating by energy release in nuclear reactions (for more details see [14]). As nuclear reactions among these nuclei we considered charge particle reactions, neutron captures and its inverse process, photo-dissociation, and β and α decay and fission. We have derived the neutron capture rates consistently for each individual mass model within the statistical model using the code MOD-Smoker [15]. The photodissociation rates were obtained from the neutron capture rates by detailed balance. For nuclei, for which the half lives are not known experimentally, we have adopted the β decay (and β delayed neutron emission) rates from the compilation of Möller *et al.* [16], which was derived from QRPA calculations on top of the FRDM mass model. We used the parametrization of Ref. [17] of the Viola-Seaborg formula to estimate the α-decay rates, which become relevant for heavy nuclei beyond lead. Finally, for nuclei with $Z > 83$, where a competition between (n, γ) and neutron induced fission can take place, we used neutron-induced reaction rates taken from [18] that are based on the FRDM mass model [6] and the Thomas-Fermi fission barriers of Myers and Swiatecki [19]. Rates for β delayed and spontaneous fission were adopted from [20]. Our fission yields were taken from the calculations of Ref. [21] which were derived using the code ABLA. This approach also gives a consistent estimate for the number of neutrons set free during the fission process.

3. Results

3.1. Evolution of the r-process abundances for the slow ejecta

For the r-process nucleosynthesis of the so-called slow ejecta we show in Fig. 1, the abundances at 3 different phases of the evolution: a) at freeze out, which we define as the moment where $R_{n/s} = 1$, b) the moment when the average timescale for β decays becomes equal to the average timescale for neutron captures, c) the final abundance, calculated at a time of 1 Gyr.

At freeze-out, the abundances show a strong odd-even staggering, moreover nuclei are pushed into the transuranium region, where the nuclei with largest abundances are located around $A \sim 280$ corresponding to the $N = 184$ shell closure, such nuclei have lower barriers and consecuently n-induced fission dominates over n-capture, and the r-process cycles to medium mass nuclei rather than producing heavier nuclei. In addition fission is also an important source of free neutrons, under such conditions the total number of neutrons produced can be divided into two components: 1) a prompt component that consists of the neutrons evaporated mainly by the highly excited fragments, and 2) a delayed component that occurs during the decay of the fragments to the instantaneous r-process path.

Once neutron captures are slower than β decays, matter decays to stability. In particular, the significant amount of matter above lead, still existing in the middle panels of figure 1, decays to finally form the lead peak. At 1 Gyr only the long-lived thorium and uranium isotopes survive. When comparing the time evolution of the third peak (from the upper to the middle to the lower panel of figure 1) one clearly notices a shift of the third peak to slightly larger mass numbers in the FRDM and HFB21 mass model but absent in the WS3 and DZ31 mass models. To understand the reason for this behavior it is important to remember that the r process operates along a path of almost constant neutron separation energy. The speed at which the r process proceeds from lighter nuclei to heavier nuclei depends on the beta-decay half-lives which increase with increasing mass number. On top of this global behavior, there are local effects induced by the presence of neutron shell closures. In the r-process path nuclei with $N \gtrsim 82$ and $N \gtrsim 126$ have the longest half-lives. At freeze-out, it corresponds to charge numbers $Z \approx 48$ and $Z \approx 70$, respectively. Alternatively, the effective r-process timescale in the region can change if the r-process path changes due to modifications of the underlying mass model. Different mass models differ substantially in their predictions in regions where there is a sudden change in the intrinsic deformation [22]. This is particularly the case around $N \sim 90$ and $N \sim 130$ where all mass models used in the present work predict a transition from spherical to deformed configurations. The particular relevance for the r process is the fact that this transition can be associated with a sudden drop in the neutron separations energies. This is the case for the FRDM mass model and its most noticeable consequence is the presence of a narrow peak around $A \sim 136$ at freeze-out (see upper panel for FRDM mass model in Fig. 1). Due to the accumulation of material in this region, the r process lasts slightly longer using the FRDM mass model when compared with the other models. The peak becomes washed out at later times due to continuous production of material in this region by fission. However, neutron captures on the fission yields are responsible for a flow of matter from the second r-process peak to heavier nuclei. This flow operates in all used mass models except in FRDM due to the fact that material is halted at $N \sim 90$.

The third peak abundance is noticeably more sensitive to nuclear masses that influence the neutron capture rates. For two of the mass models (FRDM, HFB21) the peak width is noticeably narrower than observed, the peak height is overestimated, the position shifted slightly to larger mass numbers and an abundance trough is predicted just above the peak as can be seen in Fig. 1. This is because the FRDM and HFB21 mass models predict noticeably smaller neutron separation energies than the Duflo-Zuker or the WS3 models at $N = 130$, just above the magic number $N = 126$. Thus these nuclei act as (additional) obstacles in r-process simulations. In summary, the third peak in the abundance distribution shift to higher mass numbers is caused mainly by late-time neutron captures.

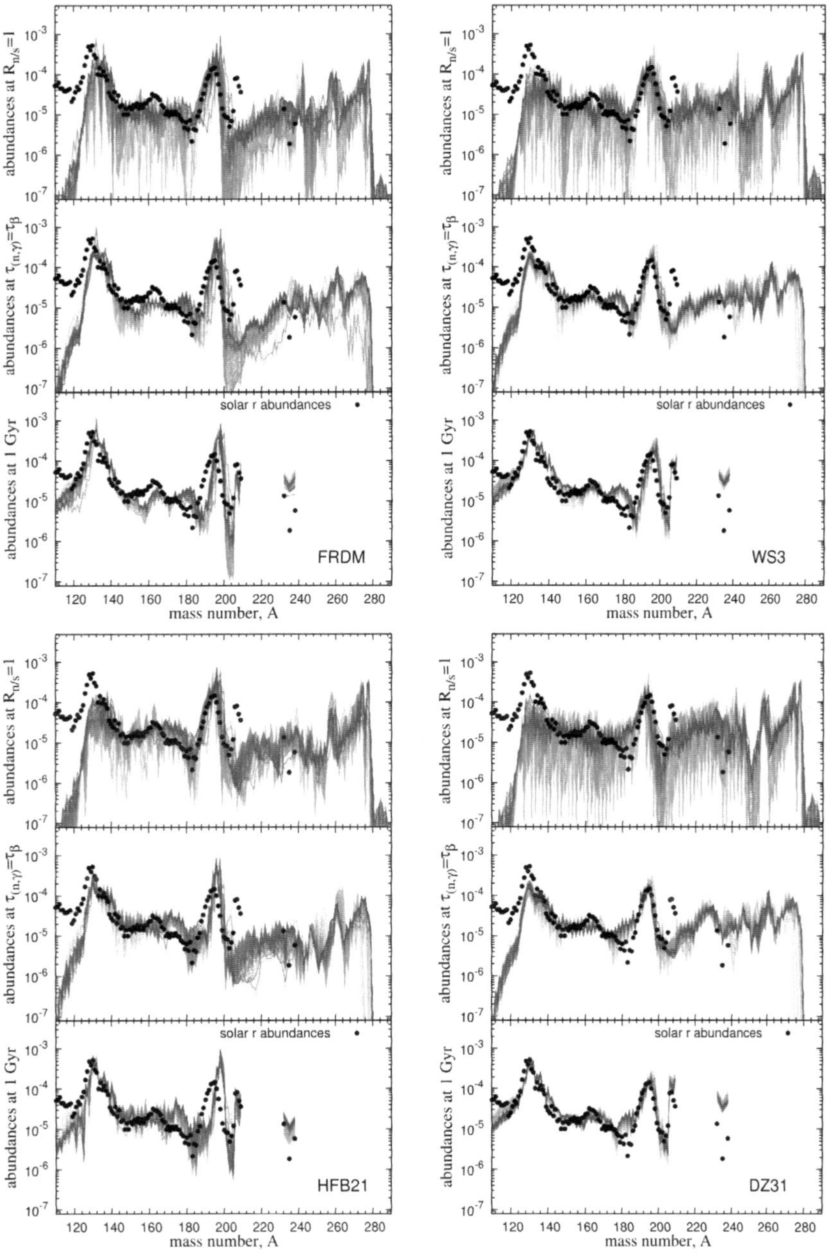

Figure 1. (Color online) R-process abundances of the slow ejecta for different mass models at different phases of the evolution. The upper panels show the abundances at $R_{n/s} = 1$. The middle panel at the time where the average timescales for beta-decay and neutron captures become identical. The lower panel shows the abundances at 1 Gyr when most of the material has already decayed to the stability.

Journal of Physics: Conference Series **730** (2016) 012018 doi:10.1088/1742-6596/730/1/012018

3.2. Evolution of the r-process abundances for fast ejecta

In Fig. 2 we show r-process abundances obtained for the fast ejecta at three different phases of the evolution and for the 4 different mass models. The top panel, for each mass model, shows the abundances before the "last" fission cycle, when the average mass number $\langle A\rangle$ reaches the final maximum. The middle panel exhibits the freeze-out abundances at $\tau_{(n,\gamma)} = \tau_\beta$ rather than by $R_{n/s} = 1$, and the lower panel shows the final r-process abundances at 1 Gyr.

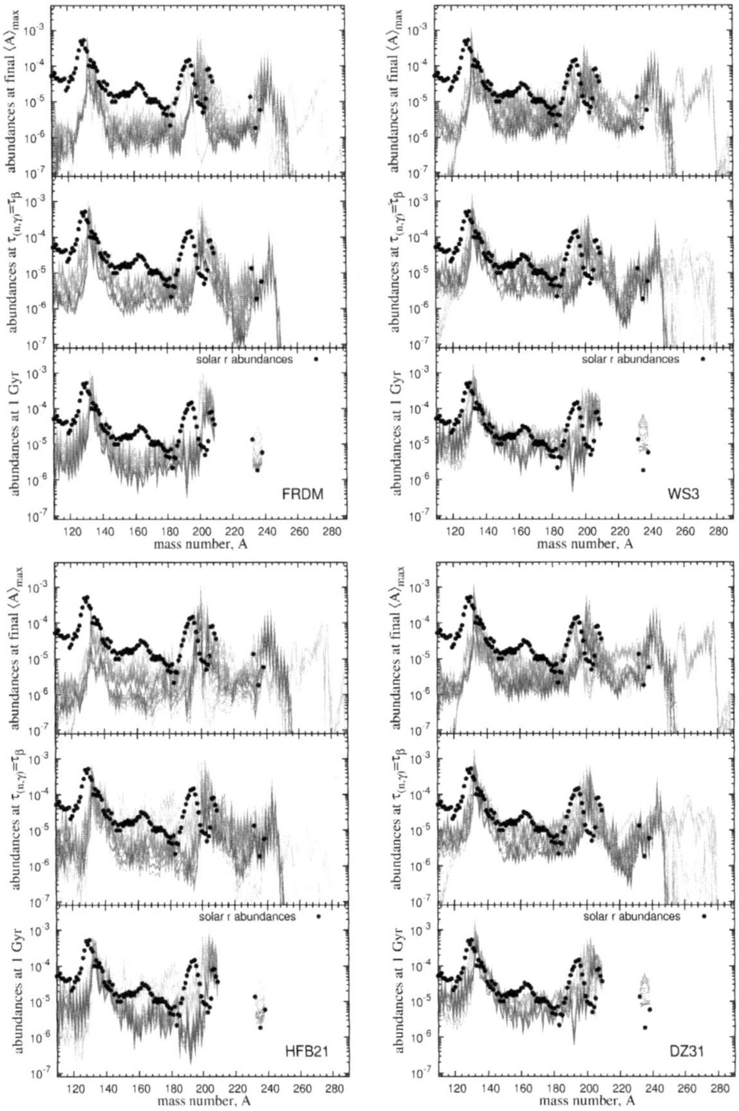

Figure 2. (Color online) R-process abundances of fast ejecta for different mass models at different phases of the evolution. The upper panels show the abundances at times before the "last" fission cycle, when the average mass number $\langle A\rangle$ reaches the final maximum. The middle panel at the time where the average timescales for beta-decay and neutron captures become identical. The lower panel shows the abundances at 1 Gyr when most of the material has already decayed to the stability.

Due to the slower neutron capture rates, the r-process path for the fast ejecta runs noticeably closer to the region of stability. This has several consequences: First, the position of third r-process peak, related to nuclei with magic neutron number $N = 126$, is shifted to larger mass numbers around $A \sim 200$. Second, under the conditions of fast expansion, and slow neutron captures, the nuclei with magic neutron numbers $N = 82$ are a noticeable obstacle for the mass flow towards heavier nuclei. Third, neutron captures are too inefficient to replenish the region of $A \sim 280$ prior to the last fission cycle for most of the trajectories. As a consequence, the subsequent decay of these heavy nuclei by fission contributes only rather modestly to the r-process abundances around the second peak at $A \sim 130$, as can be seen in the lower panels of Fig. 2. However, the decay of matter beyond the third peak, after freeze-out, fills up the abundances around lead.

In contrast to the slow ejecta, the fast ejecta exhibit a large spread in the final abundances observed between the different trajectories. This points to a very strong sensitivity to details of the astrophysical conditions and to the nuclear properties, if neutron captures are slow during the r process. In fact, the fast ejecta, encountered in our NSM scenario, resemble a nucleosynthesis process somewhat between r-process and s-process.

3.3. Robust r-process abundances

Fig. 3 shows the final abundances at times of 1 Gyr for all individual NSM trajectories and for all the mass models. Additionally, the figure exhibits mass-averaged abundances for all trajectories (red curves), for the slow trajectories (green curves), and for the fast ejecta (blue curves), respectively. To better visualize the contribution of slow and fast ejecta to the total ejected mass we have multiplied the slow and fast trajectories and their averages by the fractional contribution of each ejecta, i.e. ~ 0.9 for the slow and ~ 0.1 for the fast ejecta.

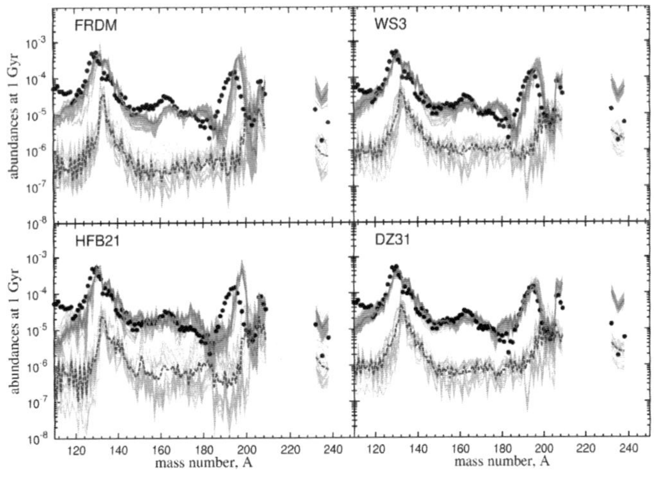

Figure 3. (Color online) Final r-process abundances at a time of 1 Gyr for the different mass models and all trajectories used in the calculations. The grey (brown) curves correspond to the abundances of the trajectories of the slow (fast) ejecta shown previously in the bottom panels of Fig. 1 and 2 but without the color gradient. The mass-averaged abundances for all trajectories (red curves), the slow ejecta (green curves), and the fast ejecta (blue curves) are also shown. The abundances for the slow and fast trajectories and their averages have been scaled by the value of their fractional contribution to the total ejecta.

As already stressed above, the most striking feature of our calculations is the fact that the final abundances for mass numbers $A > 120$ are virtually identical, for a given mass model, for all the slow ejecta, while they vary noticeably for the fast ejecta. Furthermore, the total mass-averaged abundances show the same pattern as those for the slow ejecta, as these constitute the dominating part of the ejected mass. We hence conclude that dynamical ejecta of NS mergers show a robust r-process pattern, as already concluded in refs. [3, 4, 5], provided that Y_e in the ejecta remains low, see refs [23, 24, 25]. We find that the main requirement to achieve a robust r-process pattern is that the amount of material accumulated at freeze-out in the fissioning region, $A \gtrsim 250$ is much larger than the one present in the region below the 3rd r-process peak. For the slow ejecta, this is guaranteed by the fact that the beta-decay half-lives grow with increasing mass number and by the presence of a neutron shell closure around $N = 184$. Both effects are responsible of producing a peak in the freeze-out r-process abundances around $A \sim 280$ (see upper panels Fig. 1). The material in this peak decays by fission contributing to the abundances around the 2nd r-process peak and producing a final robust r-process pattern. For the fast ejecta, the lack of material accumulated in the $A > 250$ region, due to the much slower neutron capture rates at later stage of the r-process, results in reduced impact of fission yields on the final distributions and in a much larger spread of the final abundance distributions. However, these trajectories contribute only mildly to the final mass-integrated abundances except for the region around $A = 200$.

The general features of this pattern is also independent of the mass models as it is mainly determined by fission yields. This is demonstrated in Fig. 4 where we compare the final mass-integrated abundances (at 1 Gyr) for four different mass models (FRDM, WS3, HFB21, DZ31). Although there are specific differences originating in the dependence of neutron captures on the underlying mass model, all the calculations reproduce the second and third r-process peaks reasonably well. We mention again that the peaks have different origins in our simulations: the peak around $A \sim 130$ arises from fission yields, while the peak at $A \sim 195$ reflects the $N = 126$ waiting points in the matter flow towards heavy nuclei. It is also satisfying to observe that the lead peak around $A \sim 208$ agrees reasonably well with the solar abundances. This peak is mainly produced by α decay of heavier nuclei. Finally also the abundances of the long-lived isotopes ^{232}Th and ^{238}U, which are the final product of some matter with charge numbers $Z \approx 90 - 96$, is reproduced reasonably well.

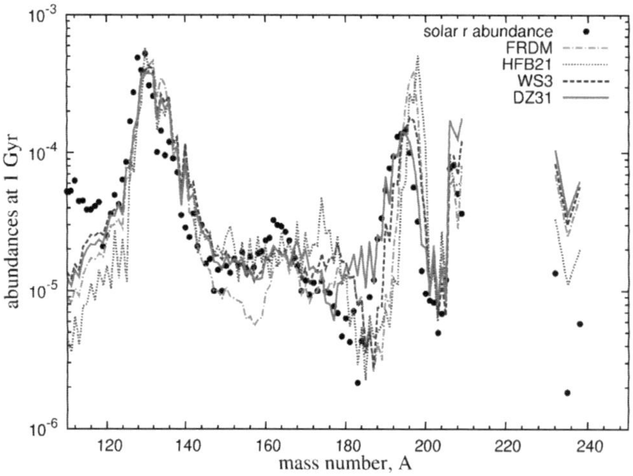

Figure 4. (Color online) Final mass-integrated abundances for all trajectories at a time of 1 Gyr for all mass models considered in this work.

4. Conclusions

We have been able to reproduce the main features of the r-process abundances (the second and third peaks, the rare-earth peak and the lead peak) reasonably well. There are modest differences in the position of the third peak and in abundance distribution just above this peak around $A \sim 205$. Here experimental work is needed to resolve these different mass predictions for the $N = 130$ nuclei.

Most of our simulations corresponding to 90% of the ejected mass, support the hypothesis that the r-process in dynamical ejecta from NSM yield rather robust abundance distribution in good agreement with the observed solar distribution for nuclei with $A \gtrsim 120$. We have shown that a requirement to achieve such a robust pattern is that at freeze-out the amount of material accumulated in the fissioning region ($A \gtrsim 250$) is much larger than the material located in the second r-process peak and above ($A \approx 120$–180). To achieve these astrophysical conditions, a sufficiently large neutron-to-seed ratio ($R_{n/s}$) is required, which, together with the fact that beta-decay half-lives along the r-process path grow with increasing mass number, guarantees the pile up of material in the fissioning region. The decay of this material by fission produces a robust r-process pattern in the region $A \approx 120$-180 that, however, depends on the used fission yields (see ref. [26]). This pattern is slightly modified by late neutron captures during the decay back to stability, which also introduces a small dependence on the astrophysical conditions.

Further extensions of the present work, need to address the impact of variations of nuclear masses for the calculation of fission and beta-decay rates.

Acknowledgments

J.J. Mendoza-Temis is a DGAPA-UNAM fellow and thanks to CONACYT project number 155663 for its support.

References

[1] Burbidge E M, Burbidge G R, Fowler W, and Hoyle F 1957 *Rev. Mod. Phys.* **29** 547
[2] Cameron A G W 1957 *Stellar Evolution, Nuclear Astrophysics, and Nucleogenesis*, Chalk River Report **CRL-41**
[3] Goriely S, Bauswein A, and Janka H T 2011 *Astrophys. J.* **738** L32
[4] Korobkin O, Rosswog S, Arcones A, and Winteler C 2012 *Mon. Not. R. Astron. Soc.* **426** 1940
[5] Bauswein A, Goriely S, and Janka H T 2013 *Astrophys. J.* **773** 78
[6] Möller P, Nix J R, Myers W D, and Swiatecki W J 1995 *At. Data Nucl. Data Tables* **59** 185
[7] Goriely S, Chamel N, and Pearson J M 2010 *Phys. Rev. C* **82** 035804
[8] Liu M, Wang N, Deng Y, and Wu X 2011 *Phys. Rev. C* **84** 014333
[9] Duflo J, and Zuker A P 1995 *Phys. Rev. C* **52** R23
[10] Mendoza-Temis J, Hirsch J G, and Zuker A P, Nuclear Physics A **843** (1), 14-36 (2010)
[11] Mumpower M, Surman R, McLaughlin G C, and Aprahamian A 2016 *Prog. Part. Nucl. Phys.* **86** 86
[12] Martin D, Arcones A, Nazarewicz W, and Olsen E 2016 *Phys. Rev. Lett.* **116** 121101
[13] Lattimer J M 2012 *Annu. Rev. Nucl. Part. Sci.* **62** 485
[14] Mendoza-Temis J J, Wu M R, Langanke K, Martínez-Pinedo G, Bauswein A, and Janka H T 2015 *Phys. Rev. C* **92** 055805
[15] Loens H P 2010 *Ph.D. thesis, TU Darmstadt* (unpublished)
[16] Mller P, Pfeiffer B, and Kratz K L 2003 *Phys. Rev. C* **67** 055802
[17] Dong T and Ren Z 2005 *Eur. Phys. J. A* **26** 69
[18] Panov I V, Korneev I Y, Rauscher T, Martínez-Pinedo G, Kelić-Heil A, Zinner N T, and Thielemann F K 2010 *Astron. Astrophys.* **513** A61
[19] Myers W D, and Swiatecki W J 1999 *Phys. Rev. C* **60** 014606
[20] Petermann I, Langanke K, Martínez-Pinedo G, Panov I V, Reinhard P G, and Thielemann F K 2012 *Eur. Phys. J. A* **48** 122
[21] Zinner N T 2007 *Ph.D. thesis, University of Aarhus* (unpublised)
[22] Arcones A and Martínez-Pinedo G 2011 *Phys. Rev. C* **83** 045809
[23] Wanajo S, Sekiguchi Y, Nishimura N, Kiuchi K, Kyutoku K, and Shibata M 2014 *Astrophys. J.* **789** L39
[24] Sekiguchi Y, Kiuchi K, Kyutoku K, and Shibata M 2015 *Phys. Rev. D* **91** 064059

[25] Goriely S, Bauswein A, Just O, Pllumbi E, and Janka H T 2015 *Mon. Not. R. Astron. Soc.* **452** 3894
[26] Goriely S, and Martínez-Pinedo G 2015 *Nucl. Phys. A* **944** 158.

Journal of Physics: Conference Series **730** (2016) 012019 doi:10.1088/1742-6596/730/1/012019

MONDE: MOmentum Neutron DEtector

P. Santa Rita[1]*, L. Acosta[1], F. Favela[1], A. Huerta[1], M. E. Ortiz[1], R. Policroniades[2] and E. Chávez[1]

[1] Instituto de Física, Universidad Nacional Autónoma de México, 04510, Cd. México., México

[2] Departamento de Aceleradores, Instituto Nacional de Investigaciones Nucleares, Carretera México-Toluca S/N, Ocoyoacac, Estado de México, 52750, México.

*Corresponding author: E-mail: pedrosarial@gmail.com

Abstract. MONDE is a large area neutron momentum detector, consisting of a 70x160x5 cm3 plastic scintillator slab surrounded by 16 photomultiplier tubes, standard NIM signal processing electronics and a CAMAC data acquisition system. In this work we present data from a characterization run using an external trigger. For that purpose, coincident gamma rays from a ^{60}Co radioactive source were used together with a NaI external detector. First results with an "external" trigger are presented.

1. Introduction.

The study of the scattering of neutrons off all kinds of targets, has relevance not only for the advancement of basic scientific knowledge, but also for its practical applications. It is a source of information for nuclear reaction mechanisms and nuclear structure. Neutron interferometry provides unique information on material structure. In industry it is used for instances to analyze large structures and humidity levels in wood. Neutrographs open a new window to study valuable art work without destroying it. Neutron scattering is also used for medical topographies. These are just a few of the many applications, it's not our purpose to mention them all [1]–[3].

When neutrons are used as probes it is often required a neutron detection system that covers a large solid angle. Some years ago, our group faced this problem [4]–[6]. In that experimental setup, a number of small neutron detectors were used to study the scattering of neutrons off Lead. Angular distributions were merely sampled and limited to a short angular range. It was realized that a very large fraction of the monochromatic neutrons scattering off the Lead target were not detected, a significant part of the effort to produce them was simply wasted. Such technical difficulties inspired the design and creation of MONDE (MOmentum Neutron DEtector) [7].

Large solid angle coverage fast neutron detection systems have been developed before, for instances: Japan [8], United States [9], Germany [10] and India [11].

In this work we use gamma rays (^{60}Co) to characterize MONDE. Taking advantage of the quasi-simultaneous emission of two gamma rays we were able to run MONDE with an external trigger.

Position of the incident tagged radiation is derived from the data with a Center of Mass algorithm.

Journal of Physics: Conference Series **730** (2016) 012019 doi:10.1088/1742-6596/730/1/012019

In section 2 all experimental details are given. Results are given in section 3 and the conclusions can be found in section 4.

2. Experimental Details.

2.1. MONDE

MONDE it's built of a single 160 x 70 x 5 cm3 sheet of BC-408 plastic scintillator, surrounded by 16 photomultiplier tubes (Thorn EMI 9903b) disposed as shown in figure 1a [12].
High-voltage for each photomultiplier is controlled independently by a LeCroy 1450 power supply to match the gain of each channel. Data acquisition is handled by a homemade program (LabView) which organizes the data in an event by event mode for off-line analysis.
MONDE can also detect, with low photoelectric efficiency, gamma rays. Taking advantage of this feature, we used a ^{60}Co source to generate an external electronic trigger. ^{60}Co beta decays to ^{60}Ni exited state (4+ at 2.5 MeV) which then decays by two sequential electromagnetic transitions (1.33 and 1.17 MeV, E2 both of them) to its ground state (0+). The intermediate state (2+) has a half-life of 9 ps [13]. The angular distribution of these gamma rays relative to each other (angular correlation) is described by the associated Legendre Polynomial of second order. Increasing the probability of emission of the two photons around 0° and 180°, relative to an isotropic distribution [14].
A general picture of MONDE is shown in figure 1b together with the simple scaffold holding the (3" x 3") NaI detector facing MONDE.

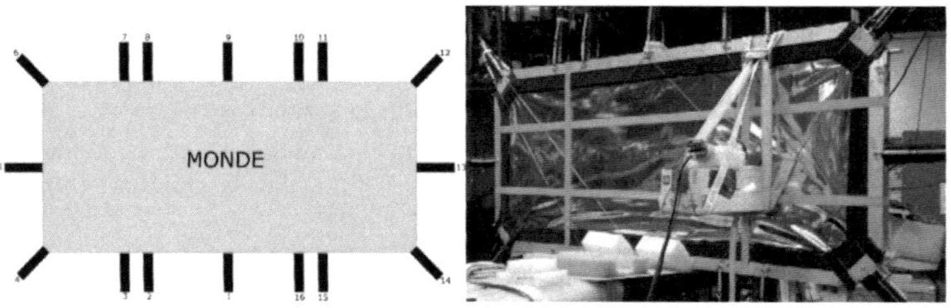

Figure 1. (a) Positions of each PMT. (b) MONDE with the external detector mounted in the scaffold.

2.2. Electronics

Figure 2 shows, schematically, the electronics used to handle all analog signals from the detectors, indicates how the logic signals are generated and used to provide a trigger for the analog to digital conversion with CAMAC modules.
When MONDE is run in standalone mode, the trigger is constructed by taking the two "t" signals of each amplifier (Mesytec MDS-8) which correspond to the average signal in each module, this two signals pass through a constant fraction discriminator (set to 50 mV) to generate two logic signals. A Logic Unit set on inclusive OR mode generates the logic signal for a valid MONDE event. An ORTEC Gate and Delay module is used to synchronize this signal with the analogic ones from the 16 PMs. Since the analog signals from the PMs are very fast, 1 μs time window width for the CAMAC ADCs is enough.

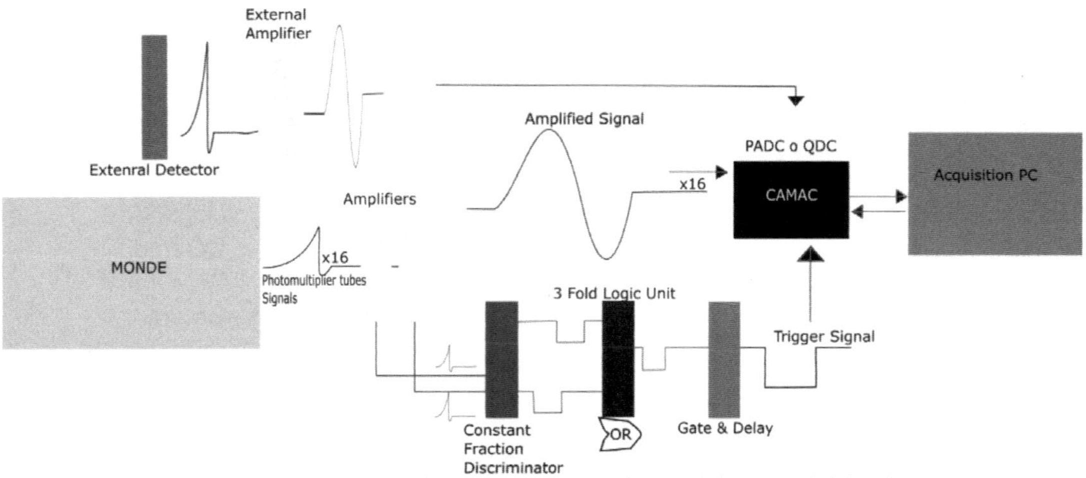

Figure 2. Electronic schematic of the signal processing and data acquisition by MONDE

To study the correlated gamma rays in MONDE and the external NaI detector (Canberra Model 802 photomultiplier tube with a 2007p preamp) we need to define a concept of "External Trigger".

Our first attempt was to generate a "logic NaI signal" from the full energy gamma peaks of ^{60}Co. However, we found that MONDE is much too fast and its analog signals were long past gone by the time we were able to construct the logic from the NaI signals. It was then decided to open wide the "waiting time" for the ADCs from 1 up to 10 μs and trigger with a logic "OR" between the MONDE and NaI signals. For count-rates below 10 kHz, this extra waiting time is not a significant problem.

The acquired data is saved in an event by event mode. Each event consists of an array of 17 integers, one corresponding to the digitalized signal from the NaI detector and 16 from each PM of MONDE.

Given the logic in which our electronic data acquisition trigger is constructed, we keep the ability to look, in an off line analysis, at MONDE and NaI "single" events, as well as conditioned ones: MONDE events given some extra condition on the NaI signals, for instances.

This is how we are set to look for "externally triggered MONDE events".

3. Results.

We will describe two data runs in this work. In the first one we took background radiation by 2 hours and in the second one a ^{60}Co was place in the vicinity of PMT number 15 (125, 0), an external NaI was placed in the scaffold, data was taken by 2 hours.
In figure 3 we superimposed spectra of all 16 PMT from the background run to show how well the gains were matched.

Figure 4 presents the spectrum of the ^{60}Co source from the NaI detector. The red area represents the photo-peak events from the two gamma rays from the Cobalt source.

In figure 5, the data from the 15th PMT detector is presented in three cases: i) background radiation (yellow). ii) When the ^{60}Co source was present (blue) and iii) events correlated with the photo-peaks in the NaI detector spectrum (red). It is worth mentioning that PM-9 showed an unexpected behavior. Even though its gain is matched and its spectrum coincides with all the others for large signals, it presented a much larger count-rate for small signals, for this reason the detector was powered off via software and ignored in the further analysis. We found out about this after data taking runs. That PM-tube has been now replaced for future experiments.

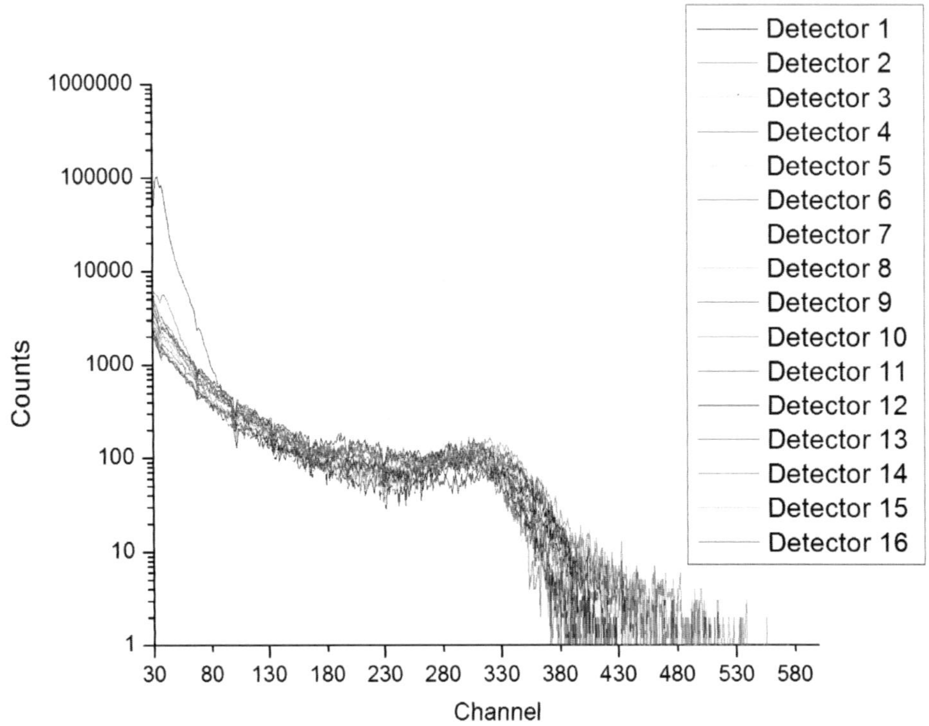

Figure 3. Background raw spectra from all the 16 PMT of MONDE superimposed.

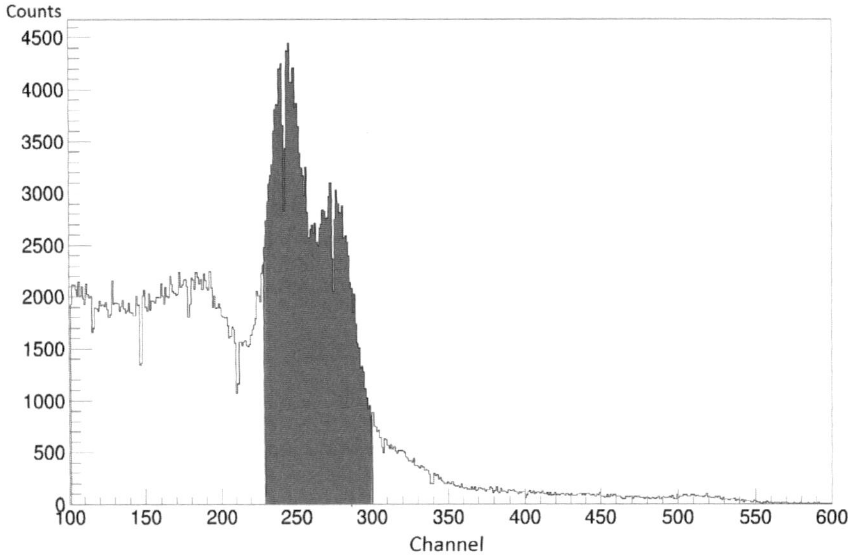

Figure 4. Spectrum of the ^{60}Co source in the NaI detector. The photo-peaks from the two gamma rays from ^{60}Co source are colored red.

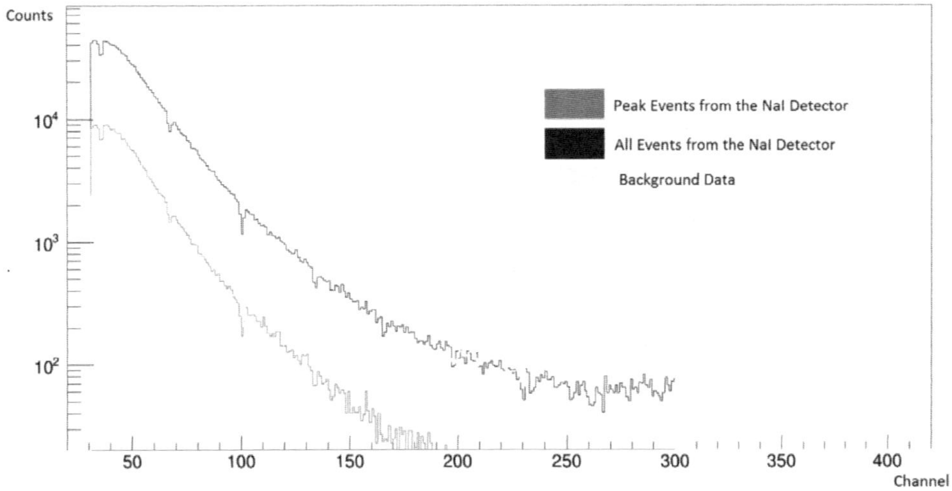

Figure 5. Superimposed spectra from the 15th PMT in different cases (see text).

Figure 5 shows a clear distinction between background and gamma ray data. It also showed that the signals produced by the gamma rays in MONDE are small. With this, we continue the analysis to reconstruct from the data, the known position of the source.

We used the C.O.M. (Centre of Mass) algorithm witch is widely used reconstructing images in ANGER cameras and imagology in medical applications [15].

Our off line analysis has been implemented in CERN ROOT (1.1 and 1.2). The program uses the signal of each photomultiplier and its known position to reconstruct the position of the spark. It is written to organize the data in highly ordered structures called "branches" this branches allowed us to manipulate the data and generate histograms to understand the data.

$$x = \frac{\Sigma_i^n x_i I_i}{\Sigma_i^n I_i} \quad \text{........ (1.1)} \qquad\qquad y = \frac{\Sigma_i^n y_i I_i}{\Sigma_i^n I_i} \quad \text{........ (1.2)}$$

In equations 1.1 and 1.2 Ii is the peak amplitude of the signal and xi, yi are the position of each PMT in MONDE.

Once we got a clear distinction between the two experiments we implement the position algorithm. A clear difference is appreciable between background in figure 6 and the ^{60}Co gamma ray source data of figure 7. The source was place in the vicinity of the PMT labeled 15 (125 cm in the X axis) where a peak is formed near that position. In figures 6 and 7, events with at least two PMT signals above a given threshold were admitted. This algorithm produces structures are artificial.

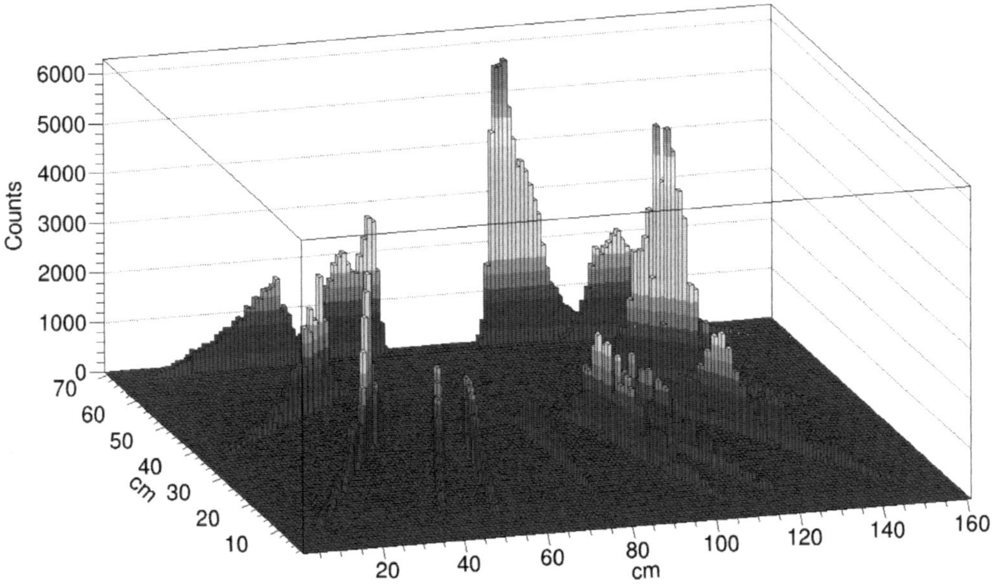

Figure 6. 3-dimensional representation (x, y, counts) of Background data using MONDE and the center of mass position reconstruction algorithm. PMT number 10 was "noisier" producing a larger count-rate than the rest.

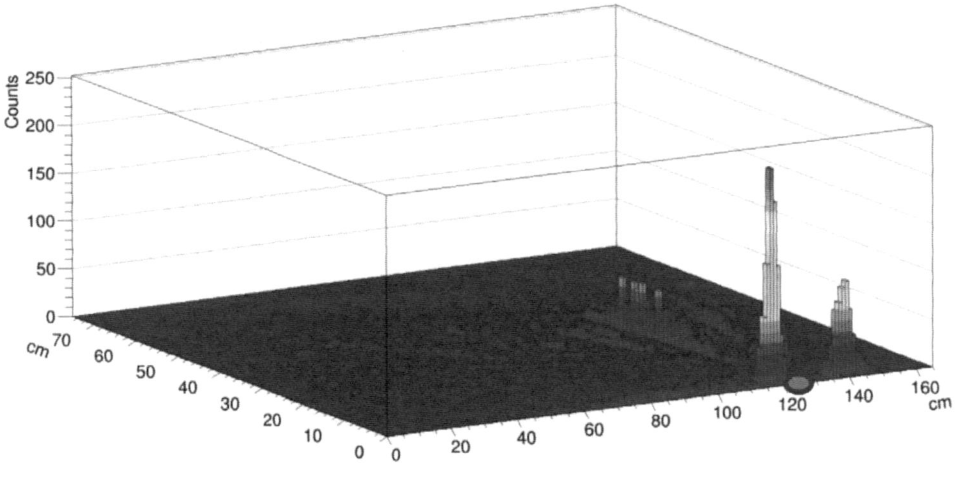

Figure 7. 3-dimensional representation of the position reconstruction of events in MONDE when the ^{60}Co gamma source is used (Blue Circle), requesting coincidence with the photo-peaks in the NaI detector.

The 10th detector shows an unusual high count making it look like a source was placed nearby. This behavior disappears when the ^{60}Co source was placed. The C.O.M. algorithm causes the line patrons in figure 6, this effect will be solved with the implementation of a new algorithm [12].

4. Conclusions

The background count-rate is important due the large volume of MONDE (typically 600 Hz). The use of an external trigger with correlated gamma rays and an external NaI detector, allowed us to select the events of interest to characterize the position response of MONDE using a simple center of mass algorithm.

This work is still in progress. In spite our efforts to carefully match the gains in all of MONDE's PMTs, the results proved that at least two out of the sixteen were working differently. Also the center of mass algorithm is too simple, a better one (PRADA) is being implemented [12], to improve the special resolution in the reconstructed position.

Position resolution is best when the source is placed in the proximities of a PMT.

Acknowledgements

This work was done with the support of CONACYT grants numbers: 51600, 82692, 123655 and DGAPAUNAM IN-118310, IN-112609, IN-105510, IN-115213, TA-100213, IN-103312.

References

[1] Schillinger B, Brunner J, and Calzada E 2006 *Phys. B Condens. Matter* **385–386** 921
[2] Hillenbach A, Engelhardt M, Abele H, and Gähler R 2005 *Nucl. Instruments Methods Phys. Res. Sect. A Accel. Spectrometers, Detect. Assoc. Equip.* **542** 116
[3] Graf B, Thillard M J, Laporte A, Le Gall J, Labat M L, and Milhaud D 1979 *Int. J. Nucl. Med. Biol.* **7** 274
[4] Chávez E, Huerta A, Ortiz M E, Rodríguez P, Favela F, Marín D, Moreno E, Murillo G, Policroniades R, Varela A, and Barrón-Palos L 2007 *Rev. Mex. Fís.* **53** 13
[5] Coello E A, Favela F, Curiel Q, and Chávez E 2012 *Rev. Mex. Fís.* **58** 198
[6] Chávez E, Rodríguez P, Huerta A, Ortiz M E, Barrón-Palos L, Favela F, Marín D, Moreno E, Murillo G, Policroniades R, and Varela A 2009 *Eur. Phys. J. A* **42** 179
[7] Huerta A, Guerrero R, Curiel Q, Huelgas J, Rodríguez P, Favela F, Marín D, Ortiz M E, Barrón-Palos L, Chávez E, Moreno E, Murillo G, Policroniades R, and Varela A 2008 *Rev. Mex. Fís.* **54** 9
[8] Kondo Y, Nakamura T, Sako T, Kawada Y, Lee G S, Kobayashi N, Ninakata R, Nishi S, Takajashi K, Tanaka R, Orr N A, Otsu H, Satou Y, and Yoneda K 2012 *RIKEN Accel. Prog. Rep.* 45 131
[9] Luther B, Baumann T, Thoennessen M, Brown J, DeYoung P, Finck J, Hinnefeld J, Howes R, Kemper K, Pancella P, Peaslee G, Rogers W, and Tabor S 2003 *Nucl. Instruments Methods Phys. Res. Sect. A Accel. Spectrometers, Detect. Assoc. Equip.* **505** 33
[10] Skeppstedt O, Roth H A, Lindström L, Wadsworth R, Hibbert I, Kelsall N, Jenkins D, Grawe H, Górska M, Moszyński M, Sujkowski Z, Wolski D, Kapusta M, Hellström M, Kalogeropoulos S, Oner D, Johnson A, Cederkäll J, Klamra W, Nyberg J, Weiszflog M, Kay J, Griffiths R, Garces Narro J, Pearson C, and Eberth J 1999 *Nucl. Instruments Methods Phys. Res. Sect. A Accel. Spectrometers, Detect. Assoc. Equip.* **421** 531
[11] Rout P C, Chakrabarty D R, Datar V M, Kumar S, Mirgule E T, Mitra A, Nanal V, and Kujur R 2009 *Nucl. Instruments Methods Phys. Res. Sect. A Accel. Spectrometers, Detect. Assoc. Equip.* **598** 526
[12] Chávez E, Barrón-Palos L, Curiel Q, Guerrero R, Huerta A, Ortiz M E, Moreno E, Murillo G, Policroniades R, and Varela A 2009 *AIP Conf. Proc.* **1099** 84
[13] Chu S Y, Nordberg H, Firestone R B, and Ekström L P 2012 *Isotope Explorer 2.23*
[14] Yang C N 1948 *Phys. Rev.* **74** 764
[15] Anger H O 1958 *Rev. Sci. Instrum.* **29** 27

Journal of Physics: Conference Series **730** (2016) 012020 doi:10.1088/1742-6596/730/1/012020

Analysis of a QCD Hamiltonian in the low energy regime.

T. Yépez-Martínez[1], D. A. Amor Quiroz[2], P. O. Hess[2] and O. Civitarese[1]

[1] Departamento de Física, Universidad Nacional La Plata, C.C.67 (1900), La Plata, Argentina
[2] Instituto de Ciencias Nucleares, UNAM, Circuito Exterior, C.U., A.P. 70-543, 04510 México, D.F., México.

E-mail: yepez@fisica.unlp.edu.ar, arturo.amor@nucleares.unam.mx, hess@nucleares.unam.mx, osvaldo.civitarese@fisica.unlp.edu.ar

Abstract. We present a QCD motivated Hamiltonian for the light quark sector. Inspired from self-consistent analysis of the Coulomb interaction, we implement an interaction of the form $(-a/r + br)$ between color sources, which already consider gluonic dynamics by the linear potential contribution. A prediagonalization of the kinetic energy term followed by the implementation of the Tamm-Dancoff method are used to obtain the eigenvalues of the Hamiltonian. A variational analysis is implemented to obtain the optimized basis for the low energy meson spectrum. The potential parameter is compared to the reported lattice string tension with relatively good agreement. The obtained energies are located close to the experimental values and further improvements are discussed.

1. Introduction

The Coulomb gauge QCD Hamiltonian is derived in [1], where the underlying interactions are dominated by the instantaneous Coulomb potential acting between color charges. In the non-Abelian theory, the potential not only couples charges but it also depends on the gluon distribution of the state in which it is calculated [2, 3, 4]. An effective Hamiltonian with the aim of describing the low energy regime of the meson spectrum should consider the dynamics of quarks and their confinement on a finite volume. Following that direction, we take the kinetic plus the mass term and the Coulomb interaction for the quarks sector from the QCD Hamiltonian. The quark kinetic term at low energy has been analyzed using the three dimensional harmonic oscillator basis [5, 6] by analytic and semi-analytic methods as well as exact diagonalization [7]. On the other hand, in [5, 6, 8] the Coulomb QCD interaction between color charge densities had been approximated by a contact interaction resulting in a second order Casimir operator of the color group. The QCD Coulomb interaction has been formally analyzed using self-consistent methods like Dyson-Schwinger equations to obtain its behavior [2, 3, 4], which in coordinate representation seems to reproduce the confining potential reported from lattice calculations [9]. Motivated by this findings, in the present work we now consider a more realistic interaction of the type Coulomb $(-\frac{a}{r})$ plus linear (br) potential to analyze the low energy spectrum of mesons when the Tamm-Dancoff method is implemented to describe any possible combination of quark-antiquarks at low energy. Finally, we discus the procedure to be taken in order to get better results compared to the reported [13] meson spectrum at low energies.

2. Coulomb gauge approach and the basic ingredients

The motivated QCD Hamiltonian, which we consider is built by the kinetic and mass terms together with a confining interaction between quark color charge densities. This Hamiltonian will resembles the gluon dynamics by introducing a linear potential.

$$H_{eff} = \int \left\{ \psi^\dagger(\mathbf{r})(-i\alpha \cdot \nabla + \beta m)\psi(\mathbf{r}) \right\} d\mathbf{r} - \frac{1}{2} \int \rho_a(\mathbf{r}) V(|\mathbf{r} - \mathbf{r}'|) \rho^a(\mathbf{r}') d\mathbf{r} d\mathbf{r}' \tag{1}$$

where $\rho^a(\mathbf{r}) = \rho^a_q(\mathbf{r}) = \psi^\dagger(\mathbf{r}) T^a \psi(\mathbf{r})$ and $\psi^\dagger(\mathbf{r}) = (\psi^\dagger_1(\mathbf{r}, \sigma, c, f), \psi^\dagger_2(\mathbf{r}, \sigma, c, f))$ with

$$\begin{aligned} \psi^\dagger_1(\mathbf{r}, \sigma, c, f) &= \sum_{Nlm\sigma cf} \mathbf{b}^\dagger_{\frac{1}{2}, Nlm, \sigma cf} R^*_{Nl}(r) Y^*_{lm}(\hat{\mathbf{r}}) \chi^\dagger_\sigma \\ \psi^\dagger_2(\mathbf{x}, \sigma, c, f) &= \sum_{Nlm\sigma cf} \mathbf{b}^\dagger_{-\frac{1}{2}, Nlm, \sigma cf} R^*_{Nl}(r) Y^*_{lm}(\hat{\mathbf{r}}) \chi^\dagger_\sigma \end{aligned} \tag{2}$$

The index $\alpha = \pm\frac{1}{2}$ denotes the upper or lower Dirac level, N refers to the number of oscillation-quanta of the orbital level in the Dirac picture and l the angular momentum. The coupled representation of angular momentum and intrinsic spin-$\frac{1}{2}$ to total spin (j) is given by

$$\mathbf{b}^\dagger_{\alpha(N,l,\frac{1}{2})j\lambda, cf} = \sum_{m\sigma} \langle lm, \frac{1}{2}\sigma | j\lambda \rangle \mathbf{b}^\dagger_{\alpha Nlm, \sigma cf} \tag{3}$$

Because of confinement, the domain of fields in Eq. (2) is expected to be restricted to a finite volume in space, where individual hadrons are located. Therefore, the eigenfunctions of the three dimensional harmonic oscillator are chosen as the single-particle orbitals.

$$R_{Nl}(r) = N_{Nl} \exp(-\frac{\gamma r^2}{2}) r^l L^{l+\frac{1}{2}}_{\frac{N-l}{2}}(\gamma r^2) \tag{4}$$

Therefore, the particles are restricted to a finite volume. Below, we show that the variational analysis provides an optimized value for the oscillator length ($\frac{1}{\sqrt{\gamma}}$).

The use of the harmonic oscillator basis for the orbital part seems to restrict the validity of the approximations to a non-relativistic theory, but this is only true when individual levels are considered to describe quark states. When the complete basis is used, any relativistic state can be expanded into the non-relativistic basis. This requires to expand the relativistic states into, in general, an infinite sum. When sufficient basis states are included, this sum can be limited in such a way that adding new terms would not modify the results.

2.1. The Hamiltonian matrix elements.

Using the fermion field quatization given in Eqs. (2-4) the Hamiltonian is expressed as

$$\begin{aligned} H_{eff} &= \sum_{\{\alpha_i N_i l_i j_i\}} K^j_{\alpha_1 N_1 l_1, \alpha_2 N_2 l_2} \left(b^\dagger_{\alpha_1(N_1 l_1 \frac{1}{2}) j_1} \cdot b^{\alpha_2(N_2 l_2 \frac{1}{2}) j_2} \right) \\ &+ \sum_{\{\alpha_i N_i l_i j_i; L\}} V^L_{\{N_i l_i j_i\}} \left[\left[b^\dagger_{\alpha_1(N_1 l_1 \frac{1}{2}) j_1} \otimes b_{\alpha_2(N_2 l_2 \frac{1}{2}) j_2} \right]^\Gamma \otimes \left[b^\dagger_{\alpha_3(N_3 l_3 \frac{1}{2}) j_3} \otimes b_{\alpha_4(N_1 l_1 \frac{1}{2}) j_4} \right]^\Gamma \right]^{\hat{0}} \end{aligned} \tag{5}$$

where $\Gamma = L,(1,1)_1,(0,0)_1$, indicating the intermediate couplings of spin, color and flavor quantum numbers. The kinetic matrix elements are given by

$$K^j_{\alpha_1 N_1 l_1, \alpha_2 N_2 l_2} = \begin{cases} k^j_{N_1 N_2} & \text{if } l_1 = j+\frac{1}{2},\ l_2 = j-\frac{1}{2},\ \alpha_1 \neq \alpha_2 \\ k^j_{N_2 N_1} & \text{if } l_1 = j-\frac{1}{2},\ l_2 = j+\frac{1}{2},\ \alpha_1 \neq \alpha_2 \\ m_0 & \text{if } (N_1 l_1) = (N_2 l_2),\ \alpha_1 = \alpha_2 = +\frac{1}{2} \\ -m_0 & \text{if } (N_1 l_1) = (N_2 l_2),\ \alpha_1 = \alpha_2 = -\frac{1}{2} \\ 0 & \text{in all other cases} \end{cases}$$

with

$$k^j_{N_1 N_2} = \sqrt{\gamma}\sqrt{\frac{N_1 - j + \frac{3}{2}}{2}}\delta_{N_2,N_1+1} + \sqrt{\gamma}\sqrt{\frac{N_1 + j + \frac{3}{2}}{2}}\delta_{N_2,N_1-1} \tag{6}$$

and the potential matrix elements

$$\begin{aligned} V^L_{\{N_i,l_i,j_i\}} &= \sqrt{8(2L+1)2\mathrm{dimF}}\frac{(-1)^L}{4(2L+1)}(-1)^{j_2+\frac{1}{2}+j_4+\frac{1}{2}}\prod_{i=1}^{4}\sqrt{(2l_i+1)(2j_i+1)} \\ &\times\langle l_1 0, l_2 0|L0\rangle\langle l_3 0, l_4 0|L0\rangle \begin{Bmatrix} j_1 & l_1 & \frac{1}{2} \\ l_2 & j_2 & L \end{Bmatrix} \begin{Bmatrix} j_3 & l_3 & \frac{1}{2} \\ l_4 & j_4 & L \end{Bmatrix} I^L_{n_i l_i} . \end{aligned} \tag{7}$$

with

$$I^L_{n_i l_i} = \int r^2 dr\ r'^2 dr' R^*_{N_1 l_1}(r) R_{N_2 l_2}(r) R^*_{N_3 l_3}(r') R_{N_4 l_4}(r') \int_{-1}^{1} d(\hat{\mathbf{r}}\cdot\hat{\mathbf{r}}') P_L(\hat{\mathbf{r}}\cdot\hat{\mathbf{r}}') V(|\mathbf{r}-\mathbf{r}'|) . \tag{8}$$

Using the confinning interaction

$$V(|\mathbf{r}-\mathbf{r}'|) = -\frac{a}{|\mathbf{r}-\mathbf{r}'|} + b|\mathbf{r}-\mathbf{r}'| , \tag{9}$$

we have found analytic expressions [15] for Eq. (8) which have been corroborated numerically.

3. The diagonalization of the kinetic energy term and the trial basis.

The kinetic term diagonalization is achieved by the method described in [7] which we resume here to consequently implement the trial basis to re-write the effective Hamiltonian.

The first step to diagonalize the kinetic energy is to consider a unitary transformation of the form

$$\boldsymbol{b}^\dagger_{\alpha(N,l,\frac{1}{2})j\lambda,cf} = \sum_{\delta k}\left(\alpha^j_{\alpha N l,\delta k}\right)^* \hat{\boldsymbol{b}}^\dagger_{\delta k j\lambda,cf} \tag{10}$$

with the following identification

$$\begin{aligned} \alpha^j_{\frac{1}{2}Nl,\frac{1}{2}k} &= \gamma^j_{Nl,k} \quad , \quad \alpha^j_{\frac{1}{2}Nl,-\frac{1}{2}k} = -\beta^j_{Nl,k} \\ \alpha^j_{-\frac{1}{2}Nl,\frac{1}{2}k} &= \beta^j_{Nl,k} \quad , \quad \alpha^j_{-\frac{1}{2}Nl,-\frac{1}{2}k} = \gamma^j_{Nl,k} \end{aligned} \tag{11}$$

leading to the transformation

$$\begin{aligned} \boldsymbol{b}^\dagger_{\frac{1}{2}(N,l,\frac{1}{2})j\lambda,cf} &= \sum_k \left(\gamma^j_{Nl,k}\boldsymbol{b}^\dagger_{kj\lambda,cf} - \beta^j_{Nl,k}\boldsymbol{d}_{kj\lambda,cf}\right) , \\ \boldsymbol{b}^\dagger_{-\frac{1}{2}(N,l,\frac{1}{2})j\lambda,cf} &= \sum_k \left(\beta^j_{Nl,k}\boldsymbol{b}^\dagger_{kj\lambda,cf} + \gamma^j_{Nl,k}\boldsymbol{d}_{kj\lambda,cf}\right) \end{aligned} \tag{12}$$

and the matrix to be diagonalized given by

$$k^j_{k_1k_2} = \sum_{\alpha_i\delta_iN_il_i} K^j_{\alpha_1N_1l_1,\alpha_2N_2l_2}\left(\alpha^j_{\alpha_1N_1l_1,\delta_1k_1}\right)^*\left(\alpha^j_{\alpha_2N_2l_2,\delta_2k_2}\right) \tag{13}$$

The corresponding eigenvalues (ϵ_{kj}) and eigenvectors allow us to re-write the Hamiltonian in the following way

$$\begin{aligned}
H_{eff} &= \sum_{kj}\epsilon_{kj}\left[\left(b^\dagger_{kj}\cdot b^{kj}\right)-\left(d_{kj}\cdot d^{\dagger\ kj}\right)\right]\\
&- \sum_{Lk_i,j_i}\left\{E^L_{k_i,j_i}\left[\left([b^\dagger_{k_1j_1}b_{k_2j_2}]^\Gamma-[d_{k_1,j_1}d^\dagger_{k_2j_2}]^\Gamma\right)\left([b^\dagger_{k_3,j_3}b_{k_4,j_4}]^\Gamma-[d_{k_3,j_3}d^\dagger_{k_4j_4}]^\Gamma\right)\right]^{\hat{0}}\right.\\
&+ F^L_{k_i,j_i}\left[\left([b^\dagger_{k_1j_1}b_{k_2j_2}]^\Gamma-[d_{k_1j_1}d^\dagger_{k_2j_2}]^\Gamma\right)\left([d_{k_3,j_3}b_{k_4j_4}]^\Gamma+[b^\dagger_{k_3j_3}d^\dagger_{k_4j_4}]^\Gamma\right)\right]^{\hat{0}}\\
&+ F'^L_{k_i,j_i}\left[\left([d_{k_1j_1}b_{k_2j_2}]^\Gamma+[b^\dagger_{k_1j_1}d^\dagger_{k_2j_2}]^\Gamma\right)\left([b^\dagger_{k_3j_3}b_{k_4j_4}]^\Gamma-[d_{k_3j_3}d^\dagger_{k_4j_4}]^\Gamma\right)\right]^{\hat{0}}\\
&+ \left.G^L_{k_i,j_i}\left[\left([d_{k_1j_1}b_{k_2j_2}]^\Gamma+[b^\dagger_{k_1j_1}d^\dagger_{k_2j_2}]^\Gamma\right)\left(d_{k_3,j_3}b_{k_4,j_4}+b^\dagger_{k_3,j_3}d^\dagger_{k_4,j_4}\right)\right]^{\hat{0}}\right\}
\end{aligned} \tag{14}$$

with

$$\begin{aligned}
E^L_{k_i,j_i} &= \sum_{N_i,l_i}V^L_{\{N_il_ij_i\}}\left[\left(\gamma^{j_1}_{N_1l_1,k_1}\gamma^{j_2}_{N_2l_2,k_2}+\beta^{j_1}_{N_1l_1,k_1}\beta^{j_2}_{N_2l_2,k_2}\right)\left(\gamma^{j_3}_{N_3l_3,k_3}\gamma^{j_4}_{N_4l_4,k_4}+\beta^{j_3}_{N_3l_3,k_3}\beta^{j_4}_{N_4l_4,k_4}\right)\right]\\
F^L_{k_i,j_i} &= \sum_{N_i,l_i}V^L_{\{N_il_ij_i\}}\left[\left(\gamma^{j_1}_{N_1l_1,k_1}\gamma^{j_2}_{N_2l_2,k_2}+\beta^{j_1}_{N_1l_1,k_1}\beta^{j_2}_{N_2l_2,k_2}\right)\left(\gamma^{j_3}_{N_3l_3,k_3}\beta^{j_4}_{N_4l_4,k_4}-\beta^{j_3}_{N_3l_3,k_3}\gamma^{j_4}_{N_4l_4,k_4}\right)\right]\\
F'^L_{k_i,j_i} &= \sum_{N_i,l_i}V^L_{\{N_il_ij_i\}}\left[\left(\gamma^{j_1}_{N_1l_1,k_1}\beta^{j_2}_{N_2l_2,k_2}-\beta^{j_1}_{N_1l_1,k_1}\gamma^{j_2}_{N_2l_2,k_2}\right)\left(\gamma^{j_3}_{N_3l_3,k_3}\gamma^{j_4}_{N_4l_4,k_4}+\beta^{j_3}_{N_3l_3,k_3}\beta^{j_4}_{N_4l_4,k_4}\right)\right]\\
G^L_{k_i,j_i} &= \sum_{N_i,l_i}V^L_{\{N_il_ij_i\}}\left[\left(\gamma^{j_1}_{N_1l_1,k_1}\beta^{j_2}_{N_2l_2,k_2}-\beta^{j_1}_{N_1l_1,k_1}\gamma^{j_2}_{N_2l_2,k_2}\right)\left(\gamma^{j_3}_{N_3l_3,k_3}\beta^{j_4}_{N_4l_4,k_4}-\beta^{j_3}_{N_3l_3,k_3}\gamma^{j_4}_{N_4l_4,k_4}\right)\right]
\end{aligned} \tag{15}$$

4. Tamm Dancoff equation and solutions.

We implement the Tamm Dancoff (TD) method [12], to define a new pair of quark anti-quark as a linear combination of the pairs $\gamma^\dagger_{12,\Gamma\mu} = \left[\boldsymbol{b}^\dagger_1\otimes\boldsymbol{d}^\dagger_2\right]^\Gamma_\mu$ where we used the short hand notation $\Gamma = (J,(0,0)_1,(\lambda_f,\lambda_f)_1)$ to denote the spin, color and flavor ($\lambda_f = 0,1$) of the pair as well as $1 = (k_1,j_1)$ and $2 = (k_2,j_2)$. Therefore, the new pair is given by

$$\tilde{\gamma}_{\kappa\Gamma\mu} = \sum_{12} C^{\kappa\Gamma}_{12}\gamma^\dagger_{12,\Gamma\mu} \tag{16}$$

which represents a mixing between all pairs formed by all possible combinations of two orbitals. The index κ labels the new pairs and it runs from zero to the number of possibilities to form pairs. The old and new vacuum coincide, because it is a linear combination of the old annihilation operators. The resulting matrix equation is of the form

$$\sum_{1'2'} M^\Gamma_{12,1'2'}C^{\kappa\Gamma}_{1'2'} = E^{\kappa\Gamma}C^{\kappa\Gamma}_{12} \tag{17}$$

with

$$M^\Gamma_{12,1'2'} = \langle\tilde{0}|\left[\gamma^{12,\Gamma\mu},\left[\boldsymbol{H},\gamma^\dagger_{1'2',\Gamma\mu}\right]\right]|\tilde{0}\rangle\ , \tag{18}$$

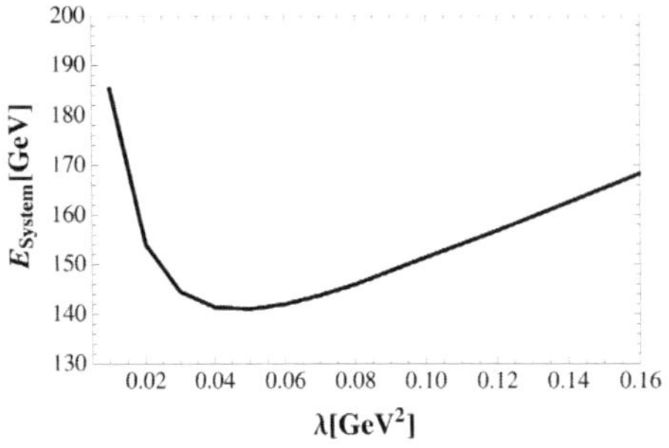

Figure 1. Variation of the energy of the system as a function of the har. osc. basis parameter.

Table 1. Tamm-Dancoff solutions for the lowest six $J = 0$ and 1 states.

E_{TD}[GeV]	J=0	J=1
E_1	0.466	0.920
E_2	0.683	0.977
E_3	0.807	1.037
E_4	0.914	1.066
E_5	0.917	1.087
E_6	1.013	1.114

where the matrix M does not depend on the magnetic numbers and it is well defined by using Eqs. (7) and (15) and numerical techniques [10].

One feature we noted is that the $-\frac{a}{r}$ interaction does not have a big influence on the results, which is due to the low energy, when long-range effects are dominant. This leaves only one parameter free (b). In order to obtain the spectrum, we have performed a variational approach on the energy of the system to optimize the harmonic oscillator parameter (γ) as it is shown on Fig. 1. This indicates a minimum on the energy of the system at about $\gamma = 0.05\text{GeV}^2$ corresponding to radius $R \approx 0.9$fm. The energy scales for the meson energy spectrum of spin $J = 0, 1$ are shown on Tab. 1. In the example presented here, we took for $b = 0.4\text{GeV}^2 \approx 1.5 \times \sigma_{lattice}$, where the lattice string tensions is typically 0.26GeV^2 [9].

In this sample calculation, we restricted to massless quarks, total single particle spin $j = \frac{1}{2}$, and take the maximal number of oscillation quanta $N = 12$. Eventhough the calculation is not yet in position to be compared with the experimental values [13], it certainly shows some improvement at the TD-level calculation compared to other TD-approach [14], Fig. 2. Showing that the finite range imposed on the domain of the fermion fields points in the right direction.

5. Conclusions and Discussion.

We have briefly presented the exact diagonalization of the kinetic energy of the QCD Hamiltonian at low energy which provides the trial basis to be used for the full diagonalization of the motivated QCD Hamiltonian. For the case of no interaction, there is a two-fold degeneracy of the states at low energy. The calculation including a confining interaction of the type $(-\frac{a}{r} + br)$ with massless quarks restricted to only $j = \frac{1}{2}$ column and a maximal number of oscillation quanta equal 12, shows no more such degeneracies. The optimization of the basis indicates that the calculation also consider a finite volume corresponding to a oscillator length of about 0.9fm.

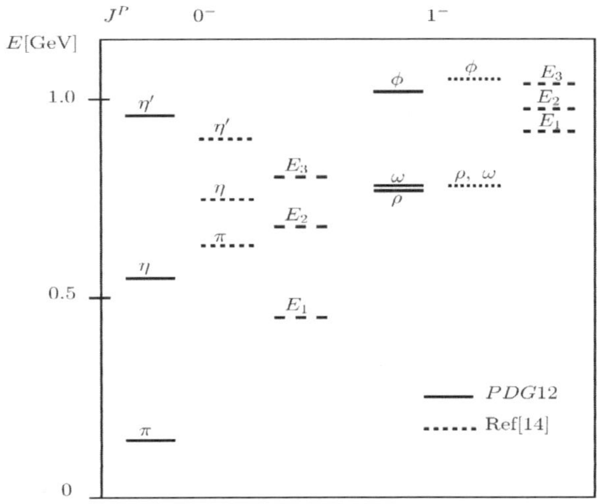

Figure 2. Tamm Dancoff energies compared to the reported in Refs. [13] and [14].

Noticing that the $-\frac{a}{r}$ contribution does not modify sensibly the results at low energy, we use only the linear interaction in order to investigate the spectrum of mesons when the Tamm-Dancoff method is implemented for each flavor-spin combination. The results indicate a patter which resembles some features of the low energy meson spectrum.

Work is in progress to include flavor symmetry breaking by using $m_{u,d} \neq m_s$ as well as the implementation of the Random Phase Approximation (RPA) in order to take into account particle-hole correlations in the vacuum of QCD, the results will be published somewhere else [15].

Acknowledgements

T. Y-M and O. C. thank to CONICET for financial support. D.A. A-Q and P. O. H thank to CONACyT for financial support.

References

[1] Lee T D 1981 *Particle Physics and Introduction to Field Theory* (Singapur, World Scientific)
[2] Szczepaniak A P and Swanson E S 2001 *Phys. Rev. D* **65** 025012
[3] Feuchter C and Reinhardt H 2004 *Phys. Rev. D* **70** 105021
[4] Epple D, Reinhardt H, and Schleifenbaum W 2007 *Phys. Rev. D* **75** 045011
[5] Yépez-Martínez T, Hess P O, Szczepaniak A P and Civitarese O 2010 *Phys. Rev. C* **81** 045204
[6] Yépez-Martínez T, Amor A, Hess P O, Szczepaniak A P and Civitarese O 2011 *Int. J. Mod. Phys. E* **20** 192
[7] Amor-Quiroz D A, Hess P O, Civitarese O and Yépez-Martínez T 2015 *J. Phys. Conf. Ser.* **639** 012014
[8] Hess P O and Szczepaniak A P 2006 *Phys. Rev. C* **73** 025201
[9] Juge, K J, Kuti J, and Morningstar C J 1998 *Nucl. Phys. B (Proc.Suppl.)* **63** 326
[10] Escher J and Draayer J P 1998 *J. Math. Phys.* **39** 5123
[11] Szczepaniak A P, Swanson E S, Ji C R, and Cotanch S R 1996 *Phys. Rev. Lett.* **76** 2011
[12] Ring P and Schuck P 1980 *The Nuclear Many Body Problem* (Heidelberg, Springer)
[13] Beringer J *etal.* (Particle Data Group) 2012 *Phys. Rev. D* **86** 010001
[14] Llanes-Estrada F J and Cotanch S R 2000 *Phys. Rev. Lett.* **84** 1102
[15] Yépez-Martínez T, Amor-Quiroz A, Hess P O and Civitarese O (to be published)